Indian Entrepreneurs in Silicon Valley

Indian Entrepreneurs in Silicon Valley

The Making of a Transnational Techno-Capitalist Class

Monica R. Biradavolu

CAMBRIA
PRESS
Amherst, New York

Requests for permission should be directed to:
permissions@cambriapress.com, or mailed to:
Cambria Press
20 Northpointe Parkway, Suite 188
Amherst, NY 14228

Library of Congress Cataloging-in-Publication Data

Biradavolu, Monica Rao, 1972-
Indian entrepreneurs in Silicon Valley : the making of a transnational techno-capitalist class / Monica R. Biradavolu.
p. cm.
Includes bibliographical references and index.
ISBN 978-1-60497-527-7 (alk. paper)
1. Alien labor, East India—California—Santa Clara Valley (Santa Clara County) 2. East Indian business enterprises—California—Santa Clara Valley (Santa Clara County) 3. East Indians—California—Santa Clara Valley (Santa Clara County) 4. Businesspeople—India. 5. High technology industries—India. 6. High technology industries—California—Santa Clara Valley (Santa Clara County) I. Title.

HD8081.E3B57 2008
338'.0408991411079473—dc22

2008012535

To my parents, B. M. Rao and B. Rajyashree

Table of Contents

List of Figures

List of Tables

Acknowledgments

This book is a culmination of an interest in sociology, which first developed when I was an undergraduate at the University of Delhi. So, to my teachers, Professor Vandana Madan, Professor Subha Sankaran, and Professor Ruby Bhatia—thank you for planting the seeds of a sociological imagination which would not have grown but for your careful nurturing and guidance.

I owe a debt I cannot repay to my thesis advisor at Duke—Professor Edward Tiryakian. This book simply would not have been written without the unfailing support, considerate criticism, and incisive insights of Ed, who has eminently played, at different times, the roles of "friend, philosopher, and guide."

The research conducted in Silicon Valley and Bangalore was made possible through the generous support provided by the International Dissertation Research Fellowship (IDRF) and the Information Technology and International Cooperation Fellowship of the Social Science Research Council (SSRC). The vigorous and stimulating discussions at the SSRC-sponsored IDRF seminar in Amsterdam and the summer fellowship at the University of California, Berkeley helped shape many of the ideas in this book.

In Silicon Valley, many thanks to my hosts, Deepa and Suresh Gajulapalli, who opened their home to me with great warmth and affection. In Bangalore, I had a home away from home because of Suhasini aunty and Ravi uncle.

To Deepa, Josh, Ahmet, Kathryn, Björn, Nilgün, and Clare—thanks to you all, I read the book with different eyes than anyone else. I see countless words of encouragement, an occasional gentle push, an even rarer reprimand scattered through these pages—invisible to anyone but me. Your amazing friendship and intellectual support sustained me throughout.

And to my parents, my brother Chenchu, and the twins who are sisters in every way but through blood, Choti and Badi, I am still waiting for a language that will express my deepest love and utmost gratitude.

For guiding me through the process with patience and dedication, I extend heartfelt thanks to the wonderful staff at Cambria Press.

Indian Entrepreneurs in Silicon Valley

Introduction

Indians in the software industry have caught the imagination of a developing nation looking for heroes to herald their standing in a global world, of observers in Silicon Valley marveling at the rise of an immigrant community, of corporations aiming to find ever new sources of labor, and of researchers wanting to empirically witness the processes of globalization. A steady, quiet, almost unnoticed flow of nearly 3 decades of skilled labor between India and the United States, beginning in the post-1965 period of immigration reform, has culminated by the turn of a new millennium into a significant force, operating at the upper echelons of the hi-tech industry and in the global center of technological innovation—Silicon Valley. It is not mere coincidence that at the same time that technologically talented Indians were making their presence felt in Silicon Valley, their brethren in Bangalore were making this unknown city in southern India the center of a developing nation's hi-tech industry. Indians, then, both within the national context of an emerging economy, and as an ethnic minority halfway across the world in the United States, offer the researcher of globalization an immediate entry into understanding how changes in a global industry impact local communities separated by geographical distances but united through new means of technological communications.

In popular perception, the term "immigrant" connotes a position of disadvantage. When the term "immigrant" is overlaid with the process of immigration from a *developing* country, the picture is one of displacement, lack of alternatives, and the attendant worries about incorporating oneself into an alien culture. Even if the migration was that of highly skilled personnel, the description was pessimistic, obvious in the term "brain drain." I am not suggesting that these images were incorrect. Indeed, the loss of skilled engineers, doctors, and technologists, trained at government expense, was not insignificant for India. However, the successful entrepreneurial activity of Indians in Silicon Valley distorts the picture. To bring the moving reel of images back into frame, one has to view India as a new, *emerging* economy, the immigration process as *transnationalism*, immigrants as a *transnational capitalist class*, and the software sector as a global industry, where Silicon Valley and Bangalore are but two inextricably interconnected nodes. This book takes such a view. Using a historical analysis, coupled with field research in Silicon Valley and Bangalore, the book highlights the changing nature of immigrant incorporation of Indians within an increasingly globalized software industry.

Chapter 1, titled "From 'The Valley of Heart's Delight' to 'Silicon Valley': The Creation of a Technical Community," attends to two issues. The first is to highlight the development of the region of Silicon Valley as a unique combination of elements, one that simultaneously nourishes technical talent and rewards an entrepreneurial spirit. The triangular relationship between Stanford University, the local electronics industry, and federal government funding rapidly transformed the region within a short span of 50 years, changing the landscape from an agricultural setting at the turn of the 19th century to the center of technological productivity by the 1950s.

The focus on the *techno*-entrepreneurial climate of the region is critical to understanding why Indian immigrants, well versed in technical skills, were able to entrench themselves deeply into the ethos of Silicon Valley—first, as a significant part of its educated workforce, and second, as the region's emerging transnational entrepreneurs. The ethos of Silicon Valley, its "culture," is one in which an environment of understandable rivalry between competing firms happily coexists with a less visible and underlying atmosphere of mutual cooperation. The Indian immigrant population of the region imbibed the tenets of this firmly grounded belief. However, replicating the culture of mainstream society to engage in entrepreneurial ventures is not what one expects to find in the literature on immigrant entrepreneurship.

Thus, secondly, the chapter grapples with the existing scholarly accounts on entrepreneurial practices of ethnic minority communities in developed societies. The case of Indian immigrants in Silicon Valley—who neither operate ventures in the traditionally low-skilled businesses that immigrants typically find themselves in, nor engage in business practices geared primarily toward a coethnic labor market relying on coethnic labor—offers a unique opportunity to study new and emerging entrepreneurship practices among highly skilled immigrants. Ignored both by the literature on international migration and by the literature on immigrant entrepreneurship, highly skilled immigrants from India do not see themselves as "ethnic" entrepreneurs using "ethnic" resources to set up and expand their business ventures. Instead, they view their actions as "being no different from the native population,"[1] in the words of an Indian software entrepreneur.

Chapter 2, "Science, Technology and the Indian State: Colonial Policies and Postcolonial Aspirations," takes a step back in the sequence of chapters in order to move a step ahead. An anomalous feature of India's development policies is that the country produces both the highest number of science and technology personnel of any developing society while simultaneously making India hold the ignominious position of the nation with the largest number of illiterates anywhere in the world. The answer to the anomaly lies in the aspirations of postcolonial India's first Prime Minister, Jawaharlal Nehru, to set India on a course of import-substitution development, with vigorous promotion of heavy industries, and a concomitant emphasis on creating government-sponsored institutions of higher learning in science and technology. The imperative behind the postcolonial embrace of Western science was the British colonial project.

Chapter 3, "Transnationalism Among Indian Software Entrepreneurs: New Modes of Immigrant Adaptation," places the rise of entrepreneurship among Indian immigrants within the emerging and growing body of literature on transnationalism. Focusing on the new ways in which immigrant entrepreneurs are engaging with their origin and destination countries, thereby altering traditional modes of immigrant incorporation, research on transnationalism has moved the field in line with issues of globalization, since immigrant business practices are increasingly being conducted in more than one local, and even national, context.

Skilled immigration from India has proceeded in three waves—the period of "brain drain," followed by a sharp increase in the influx of Indian immigrants in the form of "bodyshopping" for the IT industry in the 1990s, and finally, the current trend of outsourcing of business processes to India and the shift from permanent to temporary, short-term assignment migration. It is important to understand the three-layered stages of migration because it is the context in which entrepreneurship as a form of immigrant adaptation arose.

Following Portes, Guarnizo, and Landolt's (1999) efforts to establish transnationalism as an "emergent social field," particularly within the context of immigrant entrepreneurship, Indian immigrant business activity is viewed as a process (a) that involves a significant proportion of persons in the relevant universe; (b) where the activities of interest are not fleeting or exceptional, but now possess certain stability and resilience over time; and (c) the content of these activities is not captured by some preexisting concept, making the invention of a new term redundant. Specifically, two surveys on Indian professional immigrants, conducted more than a decade apart, highlight the shift in adaptation of Indian immigrants from an interest in professional mobility within established corporate settings to an increasing desire to engage in self-employment practices. Furthermore, these business practices are occurring within the context of the global software industry, making the activities of Indian software entrepreneurs in Silicon Valley a transnational phenomenon, thus underlining the fact that the term "brain drain" is indeed now redundant, and the phenomenon in question is better captured by the term "transnational entrepreneurship."

Chapter 4, "Global Ethnography: Methods and Sites of Research," details the research design: choice of field sites, the study population, selection of participants, the utilization of semistructured interviews and observations, and lastly, a discussion on interpretation and analysis of findings. Studies that follow events that shatter the notion of time, place, and space have to engage in field research that takes as its unit of analysis the connections that link people, hence the term "global ethnography."

Chapter 5, "Middle Class Path to Entrepreneurship: Education, Family, Networking," is an analysis of how previously risk-averse middle-class Indians overcame their reluctance to engage in business practices, a reluctance that not only stemmed from fear of a loss of a secure livelihood that salaried jobs offered but also from a negative connotation attached to money making. Indians in Silicon Valley were able to transition from professional to entrepreneur because they realized that in Silicon Valley, there was no disjuncture between educational credentials, professional standing, and the opportunity to engage in wealth-generating businesses. Silicon Valley is not merely an entrepreneurial community, but it is also a *techno-entrepreneurial* community, and it is the equal emphasis on both technical skills and desire for business start-ups that allowed technically minded Indians to consider entrepreneurship in a favorable light. Not only this, but family support also mattered, and there is an invisible presence of women—wives and mothers—in the lives of the visible entrepreneurial men. Lastly, the chapter highlights how networking has produced successive waves of Indians willing to take a chance and set up a business.

Finally, chapter 6, "Transnational Capitalist Class of Indian Immigrant Entrepreneurs," engages with the research question on conceptualizing the nature of Indian immigrant business activities in the software industry. It begins by presenting the argument put forth by scholars on why transnationalism among immigrants can, or should be, viewed as a form of subversion of the dominant forces of capitalism. When applied to high-skilled immigrants from middle-class backgrounds, factors that characterize the subject population of the study, there is little to back the claim that the entrepreneurs are engaging in a form of resistance to the larger forces that shape their choices.

The actions of Indian entrepreneurs make this group of immigrants akin to what Sklair (2001) terms the "transnational capitalist class," a class that originated in a developing country but was nourished in a developed one. As a group, the Indian entrepreneurial community in Silicon Valley expresses allegiance to the global capitalist system, over and beyond specific affiliations to nation, region, or locality. It is also a class that uses its considerable influence to take on the political project of spreading ideas of free trade, both among the Indian diaspora of Silicon Valley and among the ruling political class in India. Finally, software professionals are exerting, to complete Sklair's definition of the term, their own interests, sometimes against those of foreign interests. The chapter concludes by examining how the ideology of "transnationalism" is perceived and propagated by the actors themselves. Specifically, borrowing from Anderson's (1991) ideas on the development of the "imagined community" in the rise of nationalism, I show that likewise, transnationalism is "invented" by the Indian software community. The invention occurs through the usage of print media that serves to circulate mainstream ideas of the global capitalist system and the internationalization of professions that provides a shared language, a "jargon," through which ideas can easily transfer across national borders.

Thus, highly skilled Indian immigrants do not view themselves as victims of the global capitalist system or pawns of a larger structural chess game that they do not control. They are active and willing agents, who, having turned entrepreneurial themselves, are pushing in favor of the governing principles of global capitalism, and, true to their transnational calling, they are doing so in both their "home" and "adopted" countries.

ENDNOTE

1. Personal interview with Rajeev M., entrepreneur in Silicon Valley, January 2000.

Chapter 1

From "The Valley of Heart's Delight" to "Silicon Valley"

The Creation of a Technical Community

Silicon Valley prides itself on being, first and foremost, a unique technical community. It is an environment in which scientific and engineering talent is in high demand, creatively fostered, and amply rewarded. To be successful in the Valley, one needs to be an engineer before aspiring to be an entrepreneur; one has to be endowed with technological expertise before becoming an expert in managerial practice. The unique setting of the region and the development of its techno-entrepreneurial community allows us to understand how skilled, technically qualified immigrants, Indians among them, were welcomed into the region and able to find a footing, at first, as a highly educated workforce and, increasingly, as immigrant entrepreneurs.

The ability of members of ethnic minority groups to become entrepreneurial in a hi-tech setting is not insignificant. Traditionally, immigrants occupy niches that natives reject—exemplified by the high preponderance of immigrant-owned ethnic restaurants, gas stations, grocery stores, and other similar small-scale service-oriented businesses. Silicon Valley immigrant entrepreneurs do not follow this pattern—as they are neither in low-skilled, low-capital-intensive segments of the market, nor are they in niches wherein natives do not compete.

Nor is Silicon Valley akin to traditional immigrant spaces like Chinatown or Little India, where immigrants cloister themselves away from the larger environment in which they live. In the latter self-secluded spaces, self-employment practices are geared primarily toward catering to an ethnic clientele; establishing businesses in Silicon Valley, on the contrary, requires engagement with mainstream society. Indian immigrants in Silicon Valley not only inhabit the same space as their native counterparts, but also strongly identify with the ethic and spirit of entrepreneurship found throughout the region. Thus, a historical analysis will delineate the "making" of Silicon Valley as a starting point to grasp the seeds from which immigrants, Indians among them, were able to sow their entrepreneurial spirit.

Within a short span of half a century, this region in Northern California transformed itself from one primarily engaged in agricultural activity into the foremost center of global hi-tech productivity. The remarkable shift in the landscape, from orchards and groves to industrial parks and hi-tech firms, was simultaneously accompanied by a movement in the population away from a farming community and toward a community with a culture of innovation and entrepreneurship. Three factors enabled the creation of this hi-tech community—the influx of military dollars, the central role of Stanford University, and third, what is referred to as the "culture of entrepreneurship," a term which will be elaborated upon later in this chapter.[1]

The Pre-World War II Era: How Green Was My Valley

The name "Silicon Valley" cannot be located on maps of Northern California. What we know today as Silicon Valley is officially called the Santa Clara Valley, a 30-by-10 mile narrow strip of land, between San Francisco to the north and the Santa Cruz mountains to the south, and winding through Palo Alto, Mountain View, Sunnyvale, and Santa Clara, down to San Jose. The name "Silicon Valley" was coined in 1971 by Don Hoefler, a reporter who wrote on the history of the semiconductor industry in *Electronic News*, a weekly trade magazine.

Prior to the Second World War, and even into the late 1940s, Santa Clara Valley was primarily an agricultural region and had little experience with any industry other than agriculture. Far from being a global center for technological innovation and production, it was instead the "Valley of Heart's Delight"—a reference to the orchards of prunes, apricots, and walnuts. However, the area was negatively affected by the Great Depression. There was no demand for the specialty crops that were the mainstay of the local agricultural economy. By the time the Second World War had come to a close, growers had already begun to uproot orchards that were proving to be an unreliable source of income. Even though in 1960 there were still as many as 215 food-processing operations in the area, it was clear by the 1970s that Silicon Valley had emerged as the undisputed leader in global technological productivity (Matthews, 1999).

The decline in agriculture and the concomitant growth of manufacturing made Silicon Valley one of the fastest-growing metropolitan areas in the country. In a 30-year time span, between 1950 and 1980, the population of Santa Clara County grew from 290,000 to 1,250,000. In 1951, the Western Electronics Manufacturers Association had 20 members in Santa Clara County. After a period of merely 4 years, the number had risen to 53. By 1974, even before Silicon Valley had become a household name, approximately 800 hi-tech firms had located in the towns that make up Silicon Valley today—Sunnyvale, Mountain View, Palo Alto, and Menlo Park—employing a total of 150,000 people (Bylinsky, 1974; Findlay, 1992). The reshaping of Santa Clara Valley after 1940 also entailed a shift from the population center of San Jose to the industrial economy revolving around the research and technology center emerging at the northern end of the Valley, near Stanford University. The shift, from the "Valley of Heart's Delight" to "Silicon Valley," occurred as a result of a combination of military funding and the presence of a university that advocated close links with local industry, creating a dynamic region with a unique technical community.

In the Shadow of the Cold War: The Role of Military Spending

After his retirement in 1965, Frederick Terman, the doyen of Stanford University and often considered the "father of Silicon Valley," served as a consultant for regions that were trying to boost local economies through investment in hi-tech industries.

However, despite considerable investments by state agencies in New York, Oregon, and Texas, these attempts at imitation proved futile (Leslie, 2000). To understand why, we must look for clues in the larger geopolitical environment in which Silicon Valley was established, and at the conditions in which it flourished—aspects of which are hard to replicate in another place and at another time.

The popular image of the Valley is one in which cash-strapped innovators tinkered in their garages to create world-class technology products and highly successful commercial ventures. The phenomenal growth of companies with such humble beginnings (HP and Apple among them) attests to the veracity of the origin of these images. In fact, the "garage" has moved beyond being merely an object of popular myth, and is now preserved as a historical landmark. Outside 367 Addison Ave. in Palo Alto, a plaque entitled "The Birthplace of Silicon Valley" has the following inscription:

> This garage is the birthplace of the world's first high-technology region, "Silicon Valley." The idea for such a region originated with Dr. Frederick Terman, a Stanford University professor who encouraged his students to start up their own electronics companies in the area instead of joining established firms in the East. The first two students to follow his advice were William R. Hewlett and David Packard, who in 1938 began developing their first product, an audio oscillator, in this garage.

This California Historical Landmark #836 fits well with the ideas of free market capitalism—of unfettered entrepreneurial energy driving local economies—thereby totally discounting the role of the state and state policy as engines of change. Although clusters of small firms in Santa Clara County were instrumental in creating a rival to the large established corporations of the

East Coast, it is undeniable that Silicon Valley owes much to the legacy of the defense-driven policies of the federal government. Even HP, the "garage" leader founded in 1938, got its early start through federal monies. In a short span of 4 years, between 1941 and 1945, HP's sales went from $37,000 to $750,000 and the number of employees rose from 9 to 100, thanks to the military contracts for its electronic measuring devices and receivers (Leslie, 2000; Saxenian, 1994).

Silicon Valley was, and continues to be, a national leader in the hi-tech segment of the defense industry, including guided missiles, electronic components, and measuring and controlling devices. Silicon Valley ranks among the most "defense-dependent communities" in the nation, as table 1 shows (Gray et al., 1999).

The sharp increases in defense spending during the Second World War decisively shaped the economy of Silicon Valley. Although the benefits for the fledgling hi-tech industry on the West Coast were minimal compared to the profits accrued by rivals on the East Coast, even small contracts made a big

TABLE 1. Metropolitan-area recipients of defense contracts in the United States, 1989.

Metro Area	National Rank	Prime Contracts per Worker Employed	Ratio to U.S. Average
San Jose	1	$4,590	4.33
Washington, D.C.	2	$3,863	3.64
St. Louis	3	$3,850	3.63
Boston	4	$2,863	2.70
Cincinnati	5	$2,778	2.62
U.S. average		$1,060	1.00

Source. Gray et al. (1999).

difference. The bulk of defense contracts went to established corporations of the East Coast like RCA, GE, and Westinghouse (Abbott, 1993). However, the wartime effort created opportunities on the West Coast, and several companies were able to find creative ways of making a mark. HP and Eitel-McCullough were able to receive steady source of defense income, strengthening the electronics industry on the West Coast (Leslie, 2000).

After the end of the war, the San Jose Chamber of Commerce spent about $1 million to attract new industry to the area. The investment reaped rich dividends, with Ford, Lockheed, IBM, Syntax, Sylvania, Philco, Loral, and Raytheon setting up facilities in the Valley in the decade of the 1950s. By the mid-1970s, 82% of all federal R&D funds were directed to the aerospace and electronics sectors, and Silicon Valley's leading role in these areas ensured that there was no dearth of continuing defense funding (Malecki, 1997).

Two geographers rued the rapid changes in the Valley in 1958:

> Asphalt covered much of the best soil. Only crops producing highest returns have been able to survive. Soaring land values threaten to crowd out even those crops which grow better here than anywhere else. Soil conversion statistics indicate that 257 acres of land have been used in this county to accommodate each 1000 of population increase. (Griffin & Chatham, 1958, p. 201)

It was not only the city of San Jose and the county in which it was located that was interested in procuring Department of Defense patronage. An important piece in the puzzle is the role of Stanford University and its drive toward securing military funds for its research programs. The university, under the leadership of Frederick Terman, did more than just passively obtain federal grants; it very actively engaged in the creation of the global hi-tech center that gave the region its name.

Terman's Vision for the Valley: Bringing the Boys Back Home

When asked why she had left Philadelphia to move to San Jose in the mid-1990s, Latika G[2]., an Indian immigrant software engineer and my host during the fieldwork research in the Valley, replied,

> My husband [also an engineer] and I were graduate students in Philadelphia. We liked living in Philly. But my husband convinced me to move here. He thought this is where we would be most comfortable. We would get better jobs, find a large Indian community, and have easier access to learning about investment opportunities in hi-tech firms.[3]

The decision for Latika and her husband to leave the East Coast and move westward was not a difficult one to make; in fact, given their occupation, it was the natural and logical career-advancement move. To be upwardly mobile and take advantage of emerging opportunities in their chosen profession, they felt that they had to be in Silicon Valley. It is therefore surprising to think that only half a century earlier, Frederick Terman (1947), a Stanford University professor, was concerned about creating a hi-tech industry comparable to the one in the East and to prevent the loss of his best students to companies on the Atlantic coast.

> The West has long dreamed of an indigenous industry of sufficient magnitude to balance its agricultural resources. The war advanced these hopes and brought to the west the beginning of a great new era of industrialization. A strong

> and independent industry must, however, develop its own intellectual resources of science and technology, for industrial activity that depends upon imported brains and second-hand ideas cannot hope to be more than a vassal that pays tribute to its overlords, and is permanently condemned to an inferior competitive position. (p. 10)

An indigenous industry would provide reasons for the best engineering talent at Stanford to stay on in the region and contribute to its growth, and also attract other highly skilled professionals to migrate to the area. Within a few decades, Silicon Valley was not only witnessing migration from other parts of the United States, but the area had also emerged as a prime destination for highly skilled and educated immigrants from around the world. By 1990, more than a third (36%) of immigrant engineers from Asia reported an intention to live in the San Francisco Bay Area (Kanjanapan, 1995).

Terman's plans for Stanford encompassed not just the university campus but also the region as a whole. The son of a Stanford psychology professor, he had grown up on its campus and had attended Stanford University for his undergraduate training. He received a PhD at MIT in electrical engineering, and returned to Stanford in 1925 as professor of engineering. During the Second World War, he worked as director of a research laboratory at MIT. He was greatly impressed by the significant role played by research conducted at East Coast universities. With the intention of ensuring that Stanford would not lag behind, he returned to Stanford in 1946 with a new title, Dean of Engineering, and in his new role, he transformed the campus and its relationship with both the military and industry. Terman wanted to strengthen Stanford's engineering department in certain specialty fields, and to do this, he followed the lead of universities on the East Coast that benefited from governmental funding.

> Government-sponsored research presents Stanford, and our School of Engineering, with a wonderful opportunity if we are prepared to exploit it...We failed to take advantage of a similar opportunity presented by the research activities of the war. We are fortunate to have a second chance to retrieve our position. It is doubtful if there will ever be a third opportunity. (Leslie, 2000, p. 54)

At the time, private industry showed little interest in funding academic research, whereas federal support for basic research grew significantly, reaching approximately $140 million in 1949. Terman ensured that Stanford would be a successful recipient of federal monies. For Terman, industry was not an alternative to the government but rather an essential leg in a triangular relationship: university-government-industry. Stanford would conduct research sponsored by Washington, and industry would draw on Stanford-trained employees to develop technology for the government. Even though Terman's vision clashed with those of the then-president of Stanford University, it was Terman's vision that prevailed, and he was extremely successful in the endeavor to secure funds for the engineering programs he supervised. In 1946, Stanford's total government contracts amounted to $128,000; a decade later, due largely to Terman's efforts, total Department of Defense contracts amounted to $4.5 million. Obtaining funds was the first of many steps that led to invigorating the engineering departments at Stanford (Lowen, 1997).

The university instituted programs and invested in constructing a campus to ensure that there was a constant source of funding for Stanford's research programs, that faculty and students at the university would have direct ties with local industry, and equally importantly, created a physical space that would make it possible for a closely knit technical community to interact with one another.

Stanford, in 1953, was the first university in the country to institute an Honors Cooperative Program, through which working professionals could be updated in their specialty. The program represented the efforts of Terman to foster close and enduring ties between the university and the local technological industry, basing it on the electronics companies' need for access to research and potential employees. The program, offered through the Stanford Center for Professional Development (SCPD), continues to exist today and allows companies to send their full-time employees, on company-provided financial support, to take classes that are offered either through Internet streaming video at the students' companies or on campus. There are twelve participating engineering departments. In 1961, there were 32 companies participating in the program. Currently, the SCPD is in partnership with over 450 member companies and government organizations and delivers academic programs to more than 5,000 students annually.

Similarly, the Stanford Research Institute (SRI), an independent, nonprofit research institute, was set up in 1946, whose primary aim was to pursue practice-focused, nonprofit research that did not fit within the traditional tasks of a university. It was established with a large loan from Stanford. The institute and Stanford's engineering program each had more than $2 million in government contracts, amounts that would increase tremendously as the Cold War progressed.[4]

However, it was through the creation of Stanford Industrial Park in 1951 (now called Stanford Research Park) that the ties between the university and the region were inextricably cemented. Stanford owned 8,000 acres of land, and real-estate developers sought it with avid interest to cater to the region's steadily growing population. Meanwhile, student enrollment almost doubled following the Second World War, and the university decided to extend its campus (Luger & Goldstein, 1991). The university's founder, Leland Stanford, had stipulated that the lands could not be sold, and could only be used for educational purposes. The first plans for expansion envisioned Palo Alto, and the university, as a satellite of the city of San Francisco to the north, with a system of highways and roads that would connect the campus to the downtown area of the city. Were this plan to be followed, Stanford's environs would be strongly rooted in suburban living, and the emphasis would be on the growth of high-income housing neighborhoods that would increase the value of university lands.

The university, along with Terman, staunchly rejected the plan, criticizing it for regarding Stanford's lands as a potential subdivision rather than for enhancing the university's educational mission. Terman argued that Stanford would be better served if the lands could be put to commercial and industrial use rather than for creating residential housing. Terman wanted Palo Alto to be the center of an indigenous industry (Findlay, 1992; O'Mara, 2004).

When the park was built, the distance between the park and Stanford classrooms could be traversed on foot. Leases were granted to companies that would enhance association with Stanford faculty and students. It was not atypical for faculty to be hired as consultants of companies and for students to be hired as employees. When Varian Associates moved into the Stanford Industrial Park area, the motivation was expressed as follows: "This move would bring the company closer to old friends, ease ongoing collaborations, and improve access to graduate students in physics and engineering" (Luger & Golstein, 1991, p. 125). Sylvania set up a facility in the park and became the first participant in the Honors Cooperative Program. GE moved west in 1954, with some urging from Terman himself. As Terman explained to the engineering manager of GE's electronics division, "Some...contractual activities of a research character are difficult to handle when the cooperating parties are widely separated geographically. This would of course not be the case if the Electronics Division of G.E. found it feasible to establish a research laboratory of its own close to Stanford" (Lowen, 1997, p. 67). Sixteen of GE's top 40 scientists and engineers had been at one time either graduate students or faculty members at Stanford (Leslie, 2000). Lockheed Aerospace Company, Admiral, and Zenith followed Sylvania's and GE's lead in setting up their own subdivisions in the Industrial park.

The tradition of faculty involvement in industrial activity continues to this day. Appendix A provides an abbreviated list of some of the faculty members in the Electrical Engineering Department and their associations with local companies, demonstrating a continuing university-industry linkage. The synergies created by the Industrial park fulfilled Terman's goal of creating an indigenous industry that would stem the flow of the best talent from moving to the East coast.

Equally importantly, his vision was able to provide a communitylike feel for the companies located close to one another in the backyard of the university. The architectural layout resembled that of a university campus. In a 1954 report, administrators argued that Stanford's development presented an opportunity to create a community in which work, home, recreation and cultural life are brought together with some degree of balance and integration. Companies incorporated sports facilities, courtyards, plazas, libraries, and other campuslike features into the workplace, dismantling any perception of change when moving between the college "campus" and the work "campus." The idea was to create a space where students and faculty would be able to move from their classrooms and department buildings into the offices of companies without noticing the difference. In fact, a year after it was completed, the industrial park was a featured exhibit at the World's Fair in Brussels in 1958, and two years later, in 1960, when Charles de Gaulle visited the United States, he asked specifically to be taken on a tour of the park (Findlay, 1992). The design pioneered by Stanford Industrial Park is replicated in several regions in the world, attempting to replicate the "look" of Silicon Valley.[5]

The location of the Industrial Park was critical in developing a community of technical experts in close proximity of one another. The camaraderie that coexists with competition in a university environment was replicated within the park. This community feeling, in its turn, created a unique culture that foreshadowed Silicon Valley's emerging success as the global center for hi-tech innovation. It was also an environment that Indian immigrants, many of whom first arrived as students into the United States, found easy to adapt to.

"Culture Within the Valley": Interfirm Cooperation and Tolerance for Spin-Offs

The broad usage of the term "culture" has been criticized by anthropologists (Eriksen, 2001; Kuper, 1999; Wagner, 1981), who argue that there is a problem when the term simultaneously refers to a category used by the people researchers study (and therefore an object of study) and to an analytical concept used by researchers in their study of culture(s). The term is also criticized for being too vague to capture anything socially meaningful. Following Shore's approach, the term "culture" used here is based on observations informed by empirical fieldwork. In his study of the European Commission, Shore argued that whenever he refers to the "culture" of the EU, he uses the term to refer to how people within that organization speak of their own "culture" (Shore, 2000). Rather than use his own attributions, the author allowed the interviewees to demarcate the organizational "culture" within the Commission. By letting the observed inform the researcher, "culture" can refer to either consensus (in the form of shared meanings) or relationships of conflict (culture as a site of contestation) (Hall, 1993). In doing the same when referring to the "culture" of Silicon Valley, I therefore apply indigenous rather than analytical uses of the term.

Thus, empirical observations consistently point to two aspects that are frequently commented upon when describing the culture of Silicon Valley. The first is an unusual degree of informality and openness, even among competitors. The second is a strong sense that the region is based on a culture operating at odds with established practices on the East Coast (Saxenian, 1994). One predominant way in which Silicon Valley distinguishes itself is in the prevailing sentiment that the region was built by engineers, and not, as on the East, by lawyers, financiers, or politicians.

Lewis (2000) described the difference:

> Back East engineering had always been viewed as glorified manual labor. No one thought of Harvard or Princeton or Yale as a place you went to become an engineer. Out here engineering did not have the stigma of manual labor. Engineering was respected, maybe more than any other profession, perhaps because the original prospectors were mining engineers, and the lawyers and bankers came as an afterthought. By the mid-1950s technically minded people were aware that the region offered them a

> chance to do better for themselves than they might back East. (p. 38)

When people in the Valley socialize, they speak in technical jargon that would be little understood in other locations. Tom Wolfe (1983) wrote,

> Every year there was some place,[6] the Wagon Wheel, Chez Yvonne, Rickey's, the Roundhouse, where members of this esoteric fraternity, the young men and women of the semiconductor industry, would head after work to have a drink and gossip and brag and trade war stories about phase jitters, phantom circuits, bubble memories, pulse trains, bounceless contacts, burst modes, leapfrog tests, p-n junctions, sleeping sickness modes, slow-death episodes, RAMs, NAKs, MOSes, PCMs, PROMs, PROM blowers, PROM blasters, and teramagnitudes, meaning multiples of a million millions. (p. 362)

The community around the semiconductor industry that Wolfe described was also different from the East Coast because relationships were open and informal, even among competitors. In a write-up in *Fortune* magazine, Bylinksy (1974) wrote of how transplanted easterners would be surprised by the close-knit ties among the "technological community":

> Many of the executives in the area got to know each other as students at Stanford or as participants in local business and political affairs. The relatively close proximity of companies makes associations easier...That kind of a close-knit community where a meeting affecting, say, the semi-conductor industry brings out company presidents by the dozens were unlikely to arise in sprawling Los Angeles or in the Boston area, where companies are widely scattered. (p. 67)

The social ties in the region arose long before this location became the predominant hi-tech location of the country. As Sturgeon (2000) argued, the roots of the social interactions in the Valley can be traced to the distance, both physical and mental, that the firms in California felt from the large bureaucratic structures of the East Coast. The relatively small-sized companies in Silicon Valley also stood in stark contrast to the dominance of a few large companies on the other side of the country. Terman (1978) commented,

> I think they were "every man for himself" much more back [east]…[East Coast] manufacturers would never cooperate [on standards for vacuum tubes], partly because of the patent situation. RCA[7] dominated the patents and you couldn't leave RCA out, and if RCA was brought in, it wanted to boss everything. The group out here was involved in military production, instruments and specialized stuff, where RCA patents weren't such a dominated feature. RCA wasn't trying to build a monopoly in the instrumentation business, for example. (as quoted in Sturgeon, 2000, p. 32)

Kaplan (1999) traces the origins of the pioneers who shaped Silicon Valley and shows that most hailed from small towns on the West Coast or the Midwest—the founders of HP were from Ann Arbor, Michigan, and Puebla, Colorado. The founders of Intel were from Pescadero, California, and Denmark, Iowa. These men were acutely aware of being "outsiders" to the bureaucratic electronics firms and financial institutions of the Atlantic Seaboard and created their own "West Coast" style of doing business.

Saxenian, in comparing the firm structures of the long-established electronics companies on the East Coast with their relatively nascent counterparts on the West Coast, argues that the smaller firms in Silicon Valley created a distinct structure that separated them from the hierarchical institutional practices of the dominant companies in and around Route 128 of the Boston area. Companies in the Valley introduced what has become a distinctly "Californian" style of management, complete with casual clothing and laid-back atmosphere. In Silicon Valley, boundaries of firms were porous and information exchange was considered routine. Saxenian (1994) quotes an experienced semiconductor executive as saying, "This is a culture in which people talk to their competitors. If I had a problem in a certain area, I felt no hesitation to call another CEO and ask about the problem—even if I didn't know him. It was overwhelmingly likely that he'd answer (my question)" (p. 33).

The spirit of cooperation and cohesion among the technical community of Silicon Valley also extends to members of other related professions, who also view themselves as part of a common endeavor. Lawyers in the Valley are all well versed in laws most relevant to their client base—namely, "computer law"—particularly those surrounding protection of intellectual property. However, the growth of Silicon Valley's law firms in the early years occurred through general business counseling, and not through intellectual property practice. Clients sought legal advice on generic issues related to entrepreneurship rather than with technology law. Most local attorneys have, at one time or another, served "as dealmakers, as counselors, as gatekeepers, as proselytizers, and as matchmakers" (Suchman, 2000).

The culture of cooperation also spawned an unusual path toward entrepreneurship. Silicon Valley is famous for people leaving parent companies to set up their own businesses, an activity that, if not encouraged, is tolerated to an unusual degree. The best example is that of the company Fairchild Corporation and the "Fairchildren" that it spawned, the blue-eyed baby being the company formed when Robert Noyce and Gordon Moore left to form Intel. The experience of working at the Fairchild Corporation served as a powerful bond for many of the region's early semiconductor engineers. Saxenian (1994) wrote that of the 400 men present at a 1969 semiconductor industry conference held in the Valley, only about 25 had never worked for Fairchild.

"Fitting In": The Culture of Silicon Valley's Indian Immigrant Entrepreneurs

How do aspects of Silicon Valley's entrepreneurial culture—of cooperation and openness—translate when applied to the business practices of its immigrant entrepreneurs? First, it is important to contextualize the issue within the literature on immigrant entrepreneurship. The analysis of why, and under what circumstances, immigrants opt to use entrepreneurship as a mode of incorporation into host societies has received much scholarly attention. The interest in this research area arose because immigrants in developed nations consistently exhibit higher rates of entrepreneurship compared with the native population. A second reason is that some immigrant populations exhibit greater propensity toward entrepreneurship than others (Waldinger, Aldrich, & Ward, 1990; Yoon, 1997).

Explanations for why immigrants use entrepreneurship for economic stability focus on ethnic group characteristics or structural factors, or, more recently, a combination of the two. Within the structural explanation variant, the blocked mobility thesis contends that discrimination in the general labor market propels immigrants into self-employment (Light, 1984). Likewise, the dual labor market theory argues that immigrants are faced with a bifurcated labor market, creating a primary sector that produces jobs with secure pay, and a secondary sector typified by instability. As a result, immigrants fill secondary sector positions, including entrepreneurial ventures, in niches rejected by natives (Portes & Bach, 1985). The immigrant entrepreneurship literature is helpful in understanding the practices of less skilled immigrants who arrive without the resources—education, professional qualifications, and language skills—necessary to enable easy adaptation into a new society.

However, Indian immigrants in the hi-tech sector are not lacking in human capital. The migration path—from a developing country to a developed one—is the same as that traversed by the populations studied by the extant literature on ethnic self-employment. However, the similarity ends there. Rich in human capital, Indian immigrants in Silicon Valley possess qualifications that the general labor market desires, and upon arrival, they do not find themselves in positions that are shunned by the native population. Therefore, what we need is a fresh approach to the study of entrepreneurial practices of skilled immigrants from newly emerging economies such as India.

The question, then, is not why Indians became entrepreneurial in Silicon Valley; rather, at issue is why they became entrepreneurial when they did not have to. Indian immigrants who set up businesses in the software sector expressed attitudes similar to those discussed in the previous section on the "culture" of Silicon Valley. The attitude of cooperation fostered in the region filtered down to the ethnic entrepreneurs who emerged in the 1990s. Seemingly replicating the general Silicon Valley attitude, an Indian founder and CEO put it to me thus:

> I don't feel that I have left the KREC [Karnataka Regional Engineering College[8]] campus. We used to get together to solve problems for classes, and there was a sense that

> we were all in it together—we lived on the campus, ate together, played soccer together, went to classes together. Now, when I am heading my own show, and if I run into problems, I just call some of the same classmates, or friends of friends, who also have their own companies located right here. Of course, we are competitors, but they are more than just competitors. I think we are no different that way from people from here and how they interact.[9]

The ties that mark the relationships between the university, the entrepreneurs, and their employees in Silicon Valley closely resemble familial relationships. Within such an environment, Indians, with their own close ties of ethnicity, are able to find a space for their entrepreneurial aspirations. The previous statement articulates how Indians are able to seek assistance from other Indians in running their businesses, suggesting that ethnic bonds are a strong factor in allowing them to set up and manage their businesses.

The reliance on ethnic ties fits well with the existing literature on immigrant entrepreneurship that has documented how minority groups, when faced with blocked mobility in the general labor market, take advantage of ethnic resources to enable entrepreneurial activities (Light & Bonacich, 1988; Light & Gold, 2000; Leung, 2001; Min & Bozorgmehr, 2000; Portes, 1995; Waldinger et al., 1990). Light and Gold (2000) define ethnic resources as "features of a group that co-ethnics utilize in economic life or from which they derive economic benefit and these include identifiable skills, organizational techniques, reactive solidarity, sojourning orientation, and other characteristics based in group tradition and experience" (p. 105).

Organizational techniques that coethnic members utilize include viewing both the family unit and the wider community as a resource-mobilizing mechanism (Barrett, Jones, & McEvoy,

2001; Werbner, 1990; Westwood & Bhachu, 1998). Thus, businesses run by ethnic entrepreneurs succeed, theorists argue, because the family is as much an economic unit as a social one. Family members are expected to run the business collectively to keep labor costs down and to allow for longer operating hours. In the same way, the larger community is a source of cheap labor, should labor shortages arise.

Reactive solidarity is another kind of ethnic resource. It builds on the idea that ethnic identity is contextual and shifting. Scholars argue that it is only in the context of immigration and the geographical, cultural, and psychological dislocation arising from alien status that ethnic identity begins to take on significance for entrepreneurial behavior (Barrett, Jones, & McEvoy, 1996; Bonacich & Modell, 1980; Light & Bonacich, 1988; Portes & Zhou, 1992). This theory is useful in explaining why groups that are not entrepreneurial in their home country become so in their adopted one.

However, two aspects are strikingly different in the case of Silicon Valley's Indian entrepreneurs and their use of ethnic resources. First, Indians, do not speak of their cooperative ties with their coethnic members as a resource specific to their own group and one that distinguishes them from the "native" population. This is aptly reflected in the statement, "we are no different that way from people from here and how they interact."[10] To the contrary, the immigrants understand their actions not as being "ethnic" in origin but primarily as duplicating the prevailing cultural aspects of the region.

Equally importantly, the relationships of cooperation that exist for several Indian entrepreneurs are not restricted to the boundaries of their own ethnic group. There are several examples of Indian immigrants who have teamed up with non-Indians to establish hi-tech ventures in Silicon Valley. Sun Microsystems was cofounded by one Indian immigrant, Vinod Khosla, who collaborated with three non-Indians—Scott McNealy, Bill Joy, and Andy Bechtolstein. Similarly, Pavan Nigam (also a first-generation Indian immigrant) teamed up with Jim Clark (who founded Netscape and Silicon Graphics) to set up the online health company, Healtheon. There is no evidence of family-run enterprises among immigrant Indian-owned hi-tech businesses in Silicon Valley. Thus, Indian immigrants operating in Silicon Valley confound the expectations of immigrant entrepreneurship theorists who place emphasis on family and community. Strong family bonds do play a role, as will be shown in a later chapter; however, the nature of the work (high skilled and technical) conducted by Indian immigrants precludes the use of nonqualified members of either the family or the wider ethnic community.

Interestingly, Indians, who have maintained a presence in Silicon Valley for the past 3 decades started to engage in entrepreneurship only in the past 10 years. A long-time Indian resident of the Valley spoke of how attitudes among Indian immigrants changed, enabling them to take advantage of the growing opportunities in setting up hi-tech businesses.

> Perceptions among Indians changed from what it was in the 1970s. Two things happened—first, Indians stopped having adversarial relations with each other and started teaming with other people to start a company. Just like others who started their businesses here. Secondly, Indians started organizing. We got a group of people to organize to meet for lunch. Then, there were activities around the temple in Fremont. Having a group of Indians agreeing to build a temple itself was a big deal. Everybody wanted something different in the planning of the temple. But finally, we agreed on common goals. We also found that temples are a good way to network. You can bring

> your wives here, and your children. You can participate in various events, celebrate Indian festivals. But you can also talk about business. We cannot get into their groups, so we created our own.[11]

Once again, the speaker makes a reference to how immigrants used their ties of ethnicity to establish cooperative networks and learned the benefit of sharing common goals. Yet, the change in perceptions not only documents changing attitudes within the Indian community, but it also shows how immigrants became more adaptable to the cultural environment in which they lived and worked. Indians saw that natives were engaging in successful entrepreneurial ventures through unified action. In response, the immigrants started to organize as well.

In 1993, Asian Americans for Community Involvement conducted the largest survey of Asian professionals in Silicon Valley and found that two-thirds of those surveyed believed that advancement to managerial positions was limited by race. Interestingly, the limitations were not attributed to "racial prejudice and stereotypes" but to "old boys' networks that exclude Asians" and the "lack of role models" (AACI, 1993). There is a consistency between the statements in the previous quotation and the 1993 survey results. What Indian immigrants perceived as a weakness was not a systemic problem, that of negative racial attitudes. Instead, what kept them from realizing their full potential lay in the shortcomings within their own group—if they learned how to organize and create role models for other aspiring Indians, they might be able to circumvent lack of managerial advancement. In other words, taking a cue from their larger environment, Indians created their own "old boys' network," a resource to utilize in engaging in business activities. The idea to do so emerged as a result of engagement with mainstream society; it was not a preexisting resource of the ethnic group.

It might be argued that, faced with a lack of advancement in ranks of management, Indians sought out their ethnic group members to organize around a common cause. Thus, it was reactive solidarity that propelled them to build ethnic networks. However, the question still remains as to why this group resource was not invoked previously and why it emerged only in the past decade. Also, reactive solidarity is best used to explain why groups that are not entrepreneurial in their origin countries become so on arrival into host societies. However, Indians in Silicon Valley became entrepreneurial in the United States at the same time that similarly qualified professionals increasingly entered self-employment in the home country. To fully understand the nature of entrepreneurship among Indian immigrants, we have to appreciate their economic activity within the context of globalization, and the manner in which the software industry has allowed business activities to flourish in two geographically distant locations simultaneously.

For Indian immigrants in the Valley, the "felt" ease of setting up businesses released them from the shackles of the mentality of previous generations of highly educated migrants to the United States. The path to incorporation into the destination country was to excel in a professional capacity, rising within the ranks of management of established companies. Entrepreneurship, and the risk taking associated with it, had not been the preferred avenue of mobility. Attributing it to the culture that prevailed in his immediate surroundings, an Indian CEO in Silicon Valley, when asked about why he chose to set up his own venture, replied,

> My parents in India think it is better for me to work for a big company, even though my brother and I were in business together in India. They don't understand why I left National Semiconductor [a company in the Valley]. But

> now they think I am doing the right thing for my family. There are many people here who leave their companies and start on their own. I had an idea and decided, why not? I don't think I am any different from anybody else here who start their companies.[12]

Thus, Indians slowly began to understand that the Valley was built on an ethic of entrepreneurship, where sound technical ideas could be translated into a business initiative and that self-employment was as viable a route to economic stability as a salaried employee in a corporation.

CONCLUSION

The ability of Indian immigrant entrepreneurs to establish start-up companies was built on the ethos of Silicon Valley's techno-entrepreneurial culture. The military-university-industry linkages created a West Coast region that greatly valued technical talent. In separating itself from rivals on the East Coast, the region developed a distinct style—both in management practices within firms and relationships between firms. With educational backgrounds in the sciences, Indian immigrants found their strong foundation in technical skills to be an invaluable asset. Together with their ability to imbue the culture of cooperation that prevailed in their surrounding environs, Indian immigrants were able to find a niche for their entrepreneurial ventures. Silicon Valley, founded by outsiders to the existing entrenched bureaucratic networks of established East Coast corporations, welcomed another set of outsiders—technically qualified immigrants who were willing, ready, and able to adapt to their environment.

The next chapter shifts locations and examines the origins of the technical community of Indians. In doing so, it highlights the role of the Indian state in promoting a national policy for higher education that emphasized science and technology, and traces its origins in the colonial period that preceded it. The Indian government's abiding interest in creating a large labor pool in technical fields has had implications for both India and the United States. The Indian workforce is now among the most sought after in the global software industry, and students from India are among the most desirable in U.S. higher institutions as graduate students in science and engineering.

ENDNOTES

1. The three factors are based on the works of Findlay (1992), Gray, Bolog, Markusen, and Ock Park (1999), Hanson (1983), Kenney (2000), Mahon (1985), Malone (2002), Markusen, Hall, Campbell, and Deitrick (1991), Rogers and Larsen (1984), and Saxenian (1994).
2. All names have been changed to protect the identity of respondents.
3. Personal conversation with Latika G., Indian software engineer in Silicon Valley, January 2000.
4. SRI formally separated from the University in 1970, and changed its name to SRI International in 1977. The change came in part as a result of antiwar protests in the late 1960s. Students at Stanford University challenged the very foundations on which Terman had built the research programs at Stanford and demanded that the institute end its entire military and Southeast Asia-related research.
5. One such instance is Electronics City in Bangalore, India. It takes a few miles of driving on dusty, unpaved roads to arrive at the heart of this southern Indian city's technological center. Modeled after the park in Palo Alto, employees enjoy the same atmosphere and carefully landscaped surroundings that they would in any location in Silicon Valley. There are now about 100 companies based in Electronics City, including the biggest Indian software companies such as Infosys and Wipro, as well as international firms like HP, Siemens, and Motorola.
6. During my fieldwork year, 1999–2000, in keeping with the drive toward healthy eating, the favorite watering hole for Valley residents was Jamba Juice, which served nutrition-filled and energizing smoothies and juices. This juice company, founded in 1993 in San Luis Obispo, had several outlets around the companies in Silicon Valley. I conducted interviews there, and would oftentimes find informal sessions underway in tables around me. There are

now thousands of Jamba Juices—a popular strip mall attraction in every state.

7. RCA is the acronym for Radio Corporation of America. Founded in 1919, it became one of the leading electronics companies for much of the 20th century under the leadership of David Sarnoff. The company became defunct in 1986.
8. REC stands for Regional Engineering College, and there are 15 such colleges in India. These were established as a cooperative enterprise between the Central and individual state governments. The particular one mentioned here is in the city of Mysore, in the Southern Indian state of Karnataka.
9. Personal interview with Rajeev M., entrepreneur in Silicon Valley, January 2000.
10. See endnote 9.
11. Personal interview with Dharam A, entrepreneur and long-time resident of Silicon Valley, December 1999.
12. Personal interview with Suresh C., entrepreneur in Silicon Valley, December 1999.

Chapter 2

Science, Technology, and the Indian State

Colonial Policies and Postcolonial Aspirations

Indian immigrants have been able to forge a symbiotic relationship with Silicon Valley: The technical environment values the engineering skills the immigrants brought with them, and the migrants, in turn, have adapted to the prevailing techno-entrepreneurial culture by engaging in business activities. It is important to take a step back to examine the origins of the technically skilled Indian immigrant—in other words, to understand, as it were, whence these

Indians came. In so doing, I follow C. Wright Mills' (1970) injunction, "Unless one assumes some trans-historical theory of the nature of history, or that man in society is a non-historical entity, no social science can be assumed to transcend history. All sociology worthy of the name is 'historical sociology'" (p. 162).

In her research on the economic contributions of Silicon Valley's highly skilled immigrants, Annalee Saxenian (1999) wrote, "Silicon Valley is the home of the integrated circuit, or IC—but when local technologists claim that 'Silicon Valley is built on ICs' they refer not to chips, but to Indian and Chinese engineers" (p. 9). This remark attests to the fact that Indian engineers are an integral part of the highly educated technical community of Silicon Valley. However, it also raises the question, one with which the following pages attempt to grapple: How does a country like India, with the largest number of illiterates in the world (Mitra & Schaffer, 2004; World Bank Country Report, 2003), also produce the scientists and engineers upon whose skills the economy of the world's most advanced hi-tech center depends? The answer lies in understanding the role of the Indian state in the development of the country's technical workforce.

The transformation of India into a production center for engineering talent began at the same time that Northern California was altering its landscape to become the locus of hi-tech productivity. When Frederick Terman returned from the East Coast to implement his vision for Stanford University and its engineering programs, halfway across the world, the end of the Second World War signaled the birth of another man's hopes and aspirations. In 1947, a year after Terman's appointment as dean of engineering at Stanford, India achieved independence from nearly 200 years of colonial rule and set itself on a trajectory of nationalist development that was dictated largely by the first prime minister, Jawaharlal Nehru.

The fledgling nation, under the leadership of Nehru, embarked on a course of autarkic, import-substitution development, with vigorous promotion of heavy industries, and a concomitant emphasis on creating government-sponsored institutions of higher learning in science and technology. Laying the foundation stone of the building of the National Institute of Sciences of India (NISI)[1] in New Delhi in 1948, Nehru declared, "The scientific method is the only right method to approach life's problems; and in India today it is even more important than elsewhere because we are backward in science" (Singh, 1988, p. 64). Combining principles of socialism with ideas of rationality, Nehru's understanding of modernization and progress rested on the belief that the state had to first take an active interest in developing indigenous industries, and the desired socialistic pattern of society would follow. By the mid-1970s, the country had the third-largest cadre of scientific and technological manpower in the world, following the United States and the U.S.S.R., with more than a million personnel (Bhaneja, 1992). Although in the late 1940s, the tumultuous events in South Asia could not have seemed more unrelated to the transformation occurring in the outposts of Northern California, the connection would reveal itself some decades later. By the time Terman's vision for the university would reach maturation with the region being renamed in the 1970s as "Silicon Valley," the offspring (in the form of the highly skilled Indian immigrant) of Nehru's marriage of state-sponsored industrialization and state-subsidized higher education would begin to arrive in large numbers to form a workforce for established industries in the United States, eventually forming a significant presence in Silicon Valley.

The chapter is divided into two sections. The first section examines the British colonial project that introduced the principles upon which the future post-Independence embrace of

Western science would be based. It was under British rule that the "Anglicist" and "Orientalist"[2] debates over the medium of instruction in schools culminated in the victory of the Anglicist faction, whereby English superseded education in the vernacular (Baber, 1996). This, needless to say, had a significant impact on the future course of development of science and technology in India and for the capacity of Indian immigrants to incorporate themselves into the United States (Khadria, 2007).

The rule of empire was also, by its very nature, a rule of constant expansion. Such an impulse produced a need to classify and enumerate both the population and the lands over which the British held dominion (Appadurai, 1993). By the mid-19th century, the colonial state was actively involved in sponsoring large-scale scientific and technological projects in British India. In fact, at the time, India was the site of one of the largest state-sponsored science and technology projects undertaken anywhere, a precursor to the development of scientific knowledge in the post-Independence era. The British enumerated everything—land, fields, crops, forests, castes, tribes, and so forth, from very early in the 19th century, resulting in a need for people, especially locals, to be trained as surveyors, overseers, mechanics, and other related technical personnel (Kumar, 1995).

Lastly, education policies that began under colonialism created an English-speaking, urban Westernized elite. This section of the local Indian population benefited from the policies of the British and sometimes actively sought employment within the colonial administration. The ideas, if not the practice, of this miniscule segment of the population also held sway over the policies of the Indian National Congress (INC).[3] Even though striving for independence from the British, the orientation of the INC, at least in terms of a vision for future industrialization of the country, increasingly turned toward propagating the values and ideals that were in keeping with Britain's policies for modernizing India.

The second section will take up the thread following Independence. In many ways, the primacy given to science and technology in postcolonial India rested on the groundwork laid by the 100 years of colonial educational policy preceding Independence. There were differences, however, and the Indian government was acutely aware of how it wanted to move away from the policies of the colonizers. While the British had established a basis for a need for technological knowledge, most of it was geared toward agricultural and farming activities, with scant attention paid to the development of Indian industry or creating an environment in which engineering and research activities for locals were also considered important. Lamenting this fact, Nehru proposed broad measures to rectify the gap, and the ways in which the Indian state, under his leadership, actively encouraged the setting up of institutions devoted to science and technology will be examined later. The Indian Institutes of Technology (IITs) constitute an excellent case study to examine both the ideals upon which the Cambridge-educated Nehru built the foundations for the future of the nascent democracy, and how and why these policies became the focus of much criticism among segments of the Indian intelligentsia. The graduates of this institute are considered the most valued human capital in the country, but they have also historically been the first to leave the country as immigrants, playing an overwhelming role as Indian immigrant entrepreneurs in Silicon Valley.

Finally, the chapter concludes by highlighting the implications of state-directed policies promoting science and technology education for both India and the United States. On the one hand, this has meant the creation of a vast technologically skilled workforce that is enabling India to become the world's leading center for outsourcing of jobs in the information-technology sector (Chopra, Manasseh, & Neal, 2004). On the other hand, the United States also benefits because Indian students—the "semi-finished human capital" (Khadria, 2001, p. 45)—continue to be absorbed by U.S. universities, particularly for higher education in the fields of engineering and computer science. In the last 6 years, India has been the largest sending country of foreign students to the United States (Open Doors, 2007).

THE SCIENCE PROJECT OF THE COLONIZERS

One of the most oft-quoted reasons for why India has been able to enter the global software market is the presence of a large English-speaking population, a clear advantage when the world is increasingly using English in business and on the Internet (Arora et al., 2001; D'Costa, 2003; Patibandla & Petersen, 2002; Saxenian, 1999). However, English is spoken by only a small fraction of the population. Although there are no official statistics on the actual numbers, rarely do rough estimates exceed even 3%. However, with a country as large as India, 3% still translates to about 30 million people, a substantial absolute number.

The use of English in contemporary India is a legacy of British colonial rule, but, importantly, the debates over medium of instruction went hand in hand with the question of imparting knowledge of science and technology in the colony. As the British undertook to consolidate their rule, the issue of establishing an educational policy came to the forefront even though a Member of Parliament in Great Britain warned against such a move, arguing that "having lost America from our folly, in having allowed the establishment of schools and colleges," a similar situation might develop in India (Baber, 1996, p. 186). In the end, the practical considerations in administering the vast territory and making centralized rule possible overrode any fears about inducing mass revolt among the people.

Disseminating scientific knowledge was the key to colonial education policy. Sir Henry Maine, professor of law at the University of Cambridge and a member of the council of India, argued:

> Native thought and literature is elaborately inaccurate; it is supremely and deliberately careless of all precision in magnitude, number and time. The Indian intellect stood in need, beyond and everything else, of stricter criteria of truth. It required a treatment to harden and brace it, and scientific teaching was exactly the tonic which its infirmities called for.[4]

Similarly, Lord George Curzon, Viceroy of India at the turn of the 20th century, opined:

> We are trying to graft the science of the West on to an Eastern stem…We have raised entire sections of the community from torpor to life, and have lifted India on a higher moral plane…In proportion as we teach the masses, so we shall make their lot happier, and in proportion as they are happier so they will become more useful members of the body politic.[5]

Even though the "civilizing" mission was inherent in the motivations, as is evident from the previously mentioned writings, there were several political and economic considerations behind why scientific knowledge within the colony was deemed necessary. Early on, the British realized that they needed to become better acquainted with the territory they controlled because every accumulation of knowledge was thought to be useful to the state. To gather the required information, the British set out to "enumerate" India—and collecting data became a virtual obsession. Behind the number-generating imperative, argues Ludden (1996), was the Enlightenment rubric of objective science, the belief that "discovering" India would produce scientific truths of universal validity. How censuses, anthropological, and other statistical surveys classified, named, and mapped the territory and people of India has generated several scholarly accounts (Cohn, 1987; Dirks, 1987; Guha, 1983). Appadurai goes so far as to argue that the process of quantification became so much a part of the British political imagination that even when the scientific work being undertaken did not have any clear bureaucratic or technical utility, "the idea had become firmly implanted that a powerful state could not survive without making enumeration a central technique of social control" (1993, p. 317).

The interest in science and technology led to the establishment of a number of scientific societies, beginning with the foundation of the Asiatic Society in Calcutta in 1784. Several others followed—the Agricultural Society of India in 1820, Bombay Natural History Society in 1833, the Indian Association for the Cultivation of Science in 1876, the Indian Mathematical Society in 1907, the Indian Science Congress Association in 1914, and the Anthropological Society of Calcutta University in 1920 (Rahman, Bhargava, Qureshi, & Pruthi, 1973).

That Western science was a necessary tool in the hands of the British for effective colonization was clear to the administrators from the start. At no time did British presence in India exceed a few tens of thousands of civilians and a couple of hundred troops. Therefore, the small size of the British population in India also demanded reliance on locals for administrative and technical tasks (Raj, 2000). There was a need, as stated by an educator in 1835, "to form a class who may be interpreters between us and the millions whom we govern; a class of persons, Indian in blood and colour, but English in taste, in opinions, in morals, and in intellect" (Young, 1957, p. 724).

Once the issue of the need to set up an educational policy was agreed upon, debates arose surrounding the proper form and method of dissemination—should the teaching be in indigenous science or Western science; should the instruction be in the vernacular or in English? Even though segments of the British administrators considered indigenous languages worthy of study, the overwhelming attitude was that there was no equivalency between Western and "Oriental" languages in the study of the sciences.

Matters were finally laid to rest in 1835 when the recommendations of Thomas Macaulay[6] were put into force. The roots of the current large numbers of people trained in the Western sciences, who now form a significant section of the highly skilled immigrant population in the United States, can be traced to the colonial system of educational training. The Indian state continued to emphasize science and technology in the post-Independence period, albeit with the goal of promoting rapid industrialization.

Scientific Solutions for Social Problems: Post-Independence Development Strategy

> There cannot be any doubt that we cannot progress nationally or individually unless we profit by the lessons of science...What is science? It is a certain way of approaching problems, a certain way of seeking the truth. It is a certain empirical way whereby we get prepared to reject anything if we cannot establish or prove it. You cannot apply science in your industries keeping other departments of your life free from it. The whole scheme is unscientific. Therefore, if we want to consider

> various problems that face us as an individual and as a social group, the right way to consider those problems is to adopt the method of science. *I am a socialist because I feel that socialism is a scientific approach to the world's problems*. [Emphasis added][7]
>
> —Jawaharlal Nehru

For Nehru, India's future prospects were dependent on the twin forces of scientific technology and rapid industrialization, with a special emphasis on heavy industries. The Second Five Year Plan[8] (1956–1961) outlined specifically that,

> [i]n the long run, the rate of industrialization and the growth of national economy would depend on the increasing production of coal, electricity, iron and steel, heavy machinery, heavy chemicals, and the heavy industries generally which would increase the capacity for capital formation. One important aim is to make India independent, as quickly as possible, of foreign imports of producer goods so that the accumulation of capital would not be hampered by difficulties in securing supplies of essential producer goods from other countries. The heavy industries must, therefore, be expanded with all possible speed. (Planning Commission, Government of India, 1956, n.p.)

The consensus within the Indian National Congress in favor of setting up heavy industries did not come easily. Nehru and Mahatma Gandhi had differing views on what India's future development should rest on. Gandhi was convinced that the condition of India's rural agricultural population could be improved not by elaborate mechanization and centralized economic planning but only through adoption of decentralization, making the Indian home a center of craft and the village a primary unit of the Indian economy. Nehru, on the other hand, believed that political independence without an economic revolution was a risk to industrially backward countries since economic control would tend to pass to a limited few, resulting in glaring disparities between countries.

Both the Gandhian and Nehruvian views on the benefits of industrialization, modernization, and mechanization were a result of their respective responses to British policies on the promotion of science and technology. While Gandhi rejected the colonial praxis as not responsive to the needs of the vast majority of Indians, Nehru thought that the British had not done enough to encourage industrialization in India. It must be stated, however, that even though Nehru's ideas of practical "scientific socialism" clashed with the Gandhian notion of an ethical-spiritual type of socialism, the divergence in concepts of socioeconomic reconstruction between Nehru and Gandhi never led to any personal clashes. The two respected the value of each other's beliefs.

Not that Nehru was only interested in science. He schooled himself extensively in Western philosophy, literature, and art, and came to believe in the principles of democracy. He coupled ideas of democracy with the ideals of socialism as practiced in the Soviet Union, and called for a centrally planned economic model. An important component in the push toward heavy industrialization was the creation of a strong workforce trained in scientific fields. The Congress Party Election Manifesto of 1945, written just prior to India's independence in 1947, declared for the first time that science, and scientific research, would be one of the nation's foremost priorities. The manifesto stated that

> science, in its instrumental fields of activity, has played an ever increasing part in influencing and molding human life and will do so in even greater measure in the future. Industrial, agricultural and cultural advance, as well as national defense depends upon it. Scientific search is

> therefore a basic and essential activity of the State and should be organized and encouraged on the widest scale. (Bhaneja, 1992, p. 26)

As soon as he was elected, Nehru declared that scientific research would be a cabinet portfolio (for a Minister) to be governed directly under him as Prime Minister. The five-year planframe started under his stewardship described a four-pronged program for research—technical training; setting up of national laboratories, universities, and scientific organizations; encouragement of basic research; and the survey of natural resources, especially for prospecting of oils and minerals.

Nehru held scientists in the highest regard. Comparing the scientist and the administrator, Nehru regarded the former as superior because of his capabilities to deliver the material goods for modernization of the country. He stated that "for this reason, our conception should change and we should recognize that the engineers and scientists are far more important than the administrators, in so far as the progress of the nation is concerned" (Singh, 1988, p. 112). He further added—with foresight for how India would provide manpower to fill global shortages in the hi-tech industry:

> The world of today becomes more and more the world of scientists, engineers and technicians. In India, we are facing problems of reconstruction. In fact, the major part of our Five Year Plan depends upon the engineers of some kind or other. Thus engineers in India today have a major responsibility. I am sure they will discharge it with credit to themselves and their country. (Singh, 1988, p. 197).

It was the belief in the superiority of science, and of technological knowledge—especially as a critical factor in the development of the nation—that led Nehru to promote the establishment of the Indian Institutes of Technology (IITs). The next section uses the example of the IIT as a case study to look at the benefits as well as a critique of Nehru's policies.

Nehru's Dream Fulfilled: The Indian Institutes of Technology (IIT)

"IIT has probably produced more millionaires than any other undergraduate institution," surmised Alexander Salkever in a 1999 article. Even if the facts do not bear out this claim, the bold statement is a reminder that the name IIT carries enormous weight in every hi-tech start-up investor board room in Silicon Valley or within Fortune 500 companies. Graduates of IIT are not only represented heavily as hi-level engineers and programmers in some of the best known technology companies in the United States, but are also sought after as executives of some of the leading companies in the nation (Lewis, 2000; Salkever, 1999). The former presidents of U.S. Airways, United Airlines, and Bell Laboratories are alumni of IIT, as are the current vice president of Citigroup and the executive vice president and chief technology officer of the American Stock Exchange. Appendix B is a list of top positions held by graduates of these elite Indian institutions, either as founders of hi-tech companies or CEOs of Fortune 500 companies.

Given the impact of the alumni of IIT on Silicon Valley, it is not surprising that the golden anniversary of the founding of the first IIT, in Kharagpur, was held in Silicon Valley in January 2003. Attendees included some of the most well-known names from industry, academia, and university—the triumvirate that marked the success of Silicon Valley—Bill Gates, founder of Microsoft, John Hennesey, president of Stanford University, and Robert Blackwill, United States ambassador to India. Gates extolled IIT as an incredible institution with a worldwide impact (Murali, 2003).

The Indian Institutes of Technology were established under the guidance of the state. With unusual foresight, the government allowed and ensured that there would not be unnecessary political meddling in the running of these institutions. Nehru asserted that the IITs were expected to provide resources to scientists and technologists of the highest caliber so that the country could be technologically self-sufficient. The IITs were the brainchild of Sir Ardeshir Dalal, technocrat and member of the cabinet-in-waiting after India was granted formal independence. Dalal, like Nehru, believed that the economic future of the country rested on the development of technology. He therefore proposed the setting up of the Council of Scientific and Industrial Research at the end of the Second World War. To man those laboratories, he persuaded the U.S. government to offer hundreds of doctoral fellowships under the Technology Cooperation Mission (TCM). However, he wanted an indigenous institution to produce scientific talent, and that is how the Indian Institute of Technology was conceptualized.

In March 1946, a 22-member committee headed by N. R. Sarkar submitted an interim report titled *The Development of Higher Technical Institutions in India*. The committee recommended that world-class institutions be set up in each of the four—north, south, east and west—corners of the country. The All India Council of Technical Education (AICTE) accepted the interim report. Illustrative of the way government bureaucracy works in India, the Government of India did not pass the IIT Act until 1961, by which time five of the initial six IITs were already fully functional (Deb, 2004; Indiresan, 2003).

There are seven IITs spread across the country. IIT Kharagpur was inaugurated in 1951, the one in Bombay in 1958, IIT

Madras in 1959, Kanpur got an IIT in 1960, and Delhi in 1961. Two IITs have been added recently to this list—IIT Guwahati was established in 1995, and University of Roorkee was granted the status of an IIT in 2001.

The institutional set-up of the IITs makes them exceptional in many ways. First, in a break from the influence of the British, the IITs were modeled not after Oxford or Cambridge, but after MIT in Massachusetts. Second, foreign assistance has been a significant factor in both the establishment and the continued growth of these institutes. In the initial phase of development, there was collaboration with the U.S.S.R. (for IIT Bombay), Germany (for IIT Madras), United States (for IIT Kanpur) and United Kingdom (for IIT Delhi). Table 2 lists the assistance received in the areas of faculty training and equipment until 1979.

Third, the IITs have their own special Government of India Act, the IIT Act of 1961, which deemed the IITs to be "institutes of national importance," and equally importantly, gave complete autonomy to the institutions in the management of their affairs.

TABLE 2. Foreign assistance received by the Indian Institutes of Technology from year of establishment* until 1979.

Institution	Equipment (in US$mil)	Guest faculty from abroad		Indian faculty trained abroad	
		No.	Man months	No.	Man months
IIT Delhi	3.11	214	1114	175	2038
IIT Kanpur	1.95	120	2226	49	500
IIT Madras	5.23	75	2254	123	1300
IIT Bombay	1.15	136	2352	27	810
IIT Kharagpur	0.53	—	220	—	560

Source. Ministry of Education and Culture, Government of India Report, 1980.
*Year of establishment is mentioned on page 41.

In a significant departure from the traditional organizational set-up of universities in India, the membership on the Board does not consist of any politician or other government official. The highest office in India, the president, is the visitor to each IIT. Fourth, unlike comparable institutions in the United States, like MIT or Caltech, the government subsidizes 75% of the approximately $3,000 per year spent on every student's education. The total tuition fees are well within the reach of the middle class in India. Thus, the term "elite" is more aptly applied to the institution, not to the socioeconomic background of the student body. Lastly, entry into these schools is based solely on being tested on one examination, which is unrelated to the high school graduation exam.

The CBS news show *60 Minutes* did a segment on the IITs and captured how these institutions exemplify prestige and rank within Indian society: "Put Harvard, MIT and Princeton together and you get a sense of the status of IIT in India" (Rajghatta, 2003). In 2002, 178,000 high school students took the Joint Entrance Exam (JEE) for 3,500 slots. Compared to Harvard, which admits about 10% of total applicants, the IITs, on average, allow admittance to only 2% of total applicants.

The JEE is a rigorous examination, considered one of the toughest in the world, and one of the fairest. Unlike the practice in the United States of allocating some slots in colleges and universities to the children of alumni, all entrants into the IITs have only one obstacle: making the cut in the JEE. N. R. Narayana Murthy, founder of Indian software giant Infosys, and himself an alumnus of IIT, explained the difficulty of securing admission. "My son wanted to do computer science. But to get into the IITs he had to be in the top 200 in the country. So he chose Cornell. I also know of cases where students who couldn't get into computer science at IIT have got scholarships at MIT,

Princeton and Caltech" (Rajghatta, 2003). The exam is a two-step process. A 3-hour screening test in physics, chemistry, and mathematics is used as a first filter, and those who successfully clear this first hurdle appear for the main examination. The main exam once again tests aptitude in physics, chemistry, and mathematics—each 2 hours in duration. In 2002, 16.7% of students who appeared for the screening test qualified for the main examination. In the JEE's more than 40-year existence, no question has ever been repeated (Deb, 2004).

The efforts of the government in granting special attention to these institutions has paid off: In a ranking of Asia's best science and technology universities in 2000, the IITs take up five of the top ten spots.

However, the policies initiated by Nehru, and followed by successive governments, have also been sharply criticized. Disproportionately more public resources have been allocated to higher-education and professional programs such as engineering

TABLE 3. Asia's top-ranked science and technology schools, 2000.

Rank		School
1	(1)	Korean Advanced Institute of Science and Technology
2	(2)	Pohang Institute of Science and Technology (South Korea)
3	(6)	Indian Institute of Technology (Bombay)
4	(4)	Indian Institute of Technology (Delhi)
5	(5)	Indian Institute of Technology (Madras)
6	(3)	Tokyo Institute of Technology
7	(7)	Indian Institute of Technology (Kanpur)
8	(-)	Indian Institute of Technology (Kharagpur)
9	(8)	Nanyang Technological University (Singapore)
10	(10)	Taiwan University of Science and Technology
14	(17)	University of Roorkee, India[9]

Source. Asiaweek, 2000 Special Report.
*The numbers in parentheses refer to the 1999 ranks.

and medicine. According to a World Bank report, for almost all states in India, universalizing elementary education remains a distant goal. Furthermore, the central government offers the greatest subsidies to higher education, and even though elementary education is free for all Indian citizens, households continue to bear 43% of the costs of sending children to elementary schools (Sipahimani, 2000).

The middle class sees technical education as the sole route to economic security, and this situation is exacerbated by the political mobilization of the lower castes seeking social and economic mobility, preferably through technical education (Omvedt, 1993; Sheth & Mahajan, 1999). Moreover, the curricula within engineering (and medical) schools are lopsided, with an emphasis solely on technical training, and a disregard of instruction in the social sciences or humanities. The tradition of a well-rounded, liberal-arts education is nonexistent. Table 4 highlights the problem within the IITs—students devote very little time to subjects in the humanities, and a comparison of the curriculum in the 1960s and 1990 demonstrates that over time, the number of hours students spent gaining an education in the social or human sciences was further reduced. Such a system creates hegemony of science, and a belief in "scientific" problem solving for all social issues.

The IITs, like other educational institutions in India, operate on a policy of reserving a quota for the lower caste groups (called Scheduled Castes) and tribal peoples (called Scheduled Tribes). Fifteen percent of total seats are reserved for students who belong to Scheduled Castes, and 7.5% for those who are categorized as Scheduled Tribes. Typically, seats in the Scheduled Caste/Scheduled Tribe quotas are offered to those who get at least two-thirds of the points obtained by the last student admitted in the general category. If this reserved quota is not

TABLE 4. A comparison of total scheduled hours by disciplinary areas in the IIT curriculum (bachelor's program), 1967 and 1990.

	Kharagpur		Bombay		Madras		Kanpur		Delhi	
	1960	**1990**	**1967**	**1990**	**1966**	**1990**	**1967**	**1990**	**1967**	**1990**
Specialized										
Engineering	41.7	51.5	38.9	64.0	42.3	56.8	33.0	41.5	32.3	57.1
Sciences	24.8	18.7	32.3	15.0	23.4	17.1	23.4	26.5	31.5	13.5
Basic										
Engineering	23.4	23.2	23.9	14.0	25.8	21.1	19.4	22.8	26.4	21.4
Humanities	10.0	6.6	5.0	7.0	9.4	5.0	14.3	9.2	9.7	8.0

Source. Indiresan and Nigam (1993).

*The data for IIT Kharagpur is for 1960–1961; all others are for the year 1966–1967.

**In the 1960s, the bachelor's degree was obtained in 5 years. By the 1990s, a bachelor's degree required 4 years of education.

filled, a limited number of Scheduled Caste and Scheduled Tribe candidates who fail to meet the two-thirds criterion are admitted to a preparatory course and are required to go through a year of training. At the end of the course, these candidates are tested in physics, mathematics, chemistry, and English and, upon achieving a certain cutoff, are admitted into the first year B.Tech. programme. Close to half of seats reserved under the quota system remain vacant, and of those admitted, a significant proportion is forced to drop out because of underperformance at the end of the preparatory year (Murali, 2003). With good primary and secondary education unavailable, it is therefore not surprising that entrance into the IITs can only be achieved if one can afford private schooling or tutoring at the preuniversity level.

The IITs do not, in the end, represent a diverse student body. The typical student is not only of at least a middle-class socioeconomic background, but also urban, male, and not belonging to one of the Scheduled Castes or Scheduled Tribes. Since the entrance exam into the IITs is unrelated to how a student fares in the high school examination, a number of coaching centers have emerged across the country that train students specifically for the IIT entrance test. Urban centers have better networks to assist students in training for the tests, which also adds to the expenses on the part of the student. Women are also grossly underrepresented in the IITs. Murali (2003) offered one reason for the aberration:

> One explanation might lie in societal and parental attitudes. A widely used stereotype suggests that girls are better at rote learning and are incapable of the type of analytical and problem-solving capabilities required by the IITs. As a result, parents seem less inclined to encourage girls to pursue the IIT dream and are unwilling to invest large amounts of time, attention and energy in

> preparatory courses for their daughters. Other engineering colleges and medical schools are considered to be safer avenues. (n.p.)

The IITs also have a dismally poor record in retaining students to stay in the country. Two-thirds of the student body immigrates, primarily to the United States. Leslie Stahl began her segment on IIT, "the most important university you have never heard of" thus: "The United States imports oil from Saudi Arabia, cars from Japan, TVs from Korea and whiskey from Scotland. So what do we import from India? We import people, really smart people" (Rajghatta, 2003, n.p.)

Implications for India and the United States

The emphasis on creating institutions of higher learning in science and engineering has had implications for both India and the United States. For India, it has meant the creation of a large cadre of technical talent that is now being fully utilized, as the country has become the chief global provider of information-technology workers (Chopra et al., 2004). For U.S. universities, eager to absorb foreign students, India is the leading sending country. Moreover, these students are likely to remain in the United States after the completion of their degree courses, adding to the labor pool in knowledge-driven industries.

Science and engineering education in India has always been viewed as the path to secure employment and middle-class stability. This is all the more true in the current situation, when India is witnessing rapid employment growth in the IT sector both as a result of the increasing presence of several multinational corporations and the outsourcing of jobs from developed nations (Mooney & Neelakantan, 2004). To keep pace with the growing demand for IT workers, Indian educational institutions continue to build on the foundation laid by the policies undertaken since independence. In 2000–2001, there were 254 universities in the country, up from 184 just 10 years previously (Khadria, 2004a). Table 5 shows the steady growth of enrolment in higher education in the fields of engineering and technology in the past decade. The enrolment numbers include three principal levels of qualification within the higher education system in India: bachelor's, master's, and doctoral degrees.

Table 6 provides an estimate of the growth in the last decade in the total stock of engineering personnel in the country, as the numbers are broken down for both the engineering-degree as well as engineering-diploma holder. There is a combined total of 2.5

TABLE 5. Enrolments in higher education in engineering and technology in India, 1991–2000 (in thousands).

Year	1991–92	1995–96	1996–97	1997–98	1998–99	1999–00
No.	258	316	331	347	363	389

Source. Khadria (2004a, p. 14).

TABLE 6. Estimated stock of engineering personnel in India (in thousands).

Year	Engineering Degree Holders	Engineering Diploma Holders
1991	547	874
1996	705	1,138
1998	798	1,256
1999	846	1,313
2000	970	1,456

Source. Khadria (2004a, p. 16).

million engineering-degree and engineering-diploma holders, as of the year 2000. Despite the large numbers, software companies in India are facing worker shortages, with top companies reporting high attrition rates (Aiyar, 2004) because of the enormous growth of the IT sector—between 2003 and 2004, the industry grew by 54% (NASSCOM, 2004a).

With the growth of India's pool of qualified personnel, there has been a concomitant rise in the number of Indian students who are leaving to pursue higher education in the United States. In 2001, for the first time, India became the largest exporter of foreign students to U.S. universities. In one year, between academic years 2000–2001 and 2001–2002, the number of students from India jumped 22.3%, compared with a growth of 5.5% for students from China, the previously top-ranked sending country (Wheeler, 2002). Seventy eight percent of Indian students who migrate overseas to pursue higher education choose the United States. More pertinent for the focus here on scientific and technological education, India also accounted for a high share of foreign-born doctorates who received their PhDs in U.S. universities—a total of 16%, second only to China (20%) (Khadria, 2004a). Furthermore, the National Science Foundation (NSF) reports that of the total foreign-born science and engineering faculty in the United States, the largest numbers from a single country are from India, followed by China, the United Kingdom, Taiwan, Germany, South Korea, Greece, and Japan (National Science Board, 2000).

However, even as the United States continues to be the traditional leader in attracting foreign students, recent trends indicate a nationwide slump, largely because of heightened security measures in the aftermath of 9/11. Real and perceived problems in obtaining visas are hampering student interest in considering the United States as an option for higher education (Dillon,

2004; Open Doors, 2004). Indian students continue to choose U.S. universities, but increasingly, other nations are competing as viable alternatives. The United Kingdom, Canada, Australia, and New Zealand are launching new scholarship programs and citizenship schemes to induce students from India and other nations to pursue education in their countries (Khadria, 2001).

CONCLUSION

Educational policies, beginning in the colonial period, set India on a course of creating a talented labor pool trained in the best practices of technical and engineering skills. The United States has been a beneficiary of India's emphasis on higher education in technological fields. Through targeted immigration policies, the United States has traditionally relied on labor trained in overseas locations to fill labor shortages within its own borders. India has been among the leading sending countries of professionally qualified workers, many of whom arrive first as students and choose to stay on. In turn, this country rewards skilled immigrants, who can avail of the employment opportunities in remunerative knowledge-based industries. However, in the current scenario, the employment opportunity for skilled workers in India is growing at the same time that white-collar jobs in the United States are being outsourced. In order to remain internationally competitive, U.S. policies have to find a way of continuing to attract immigrants who can now either find opportunities in their home countries or who find the process of legal entry to the United States to be too daunting.

Endnotes

1. In 1970, the institute was renamed as the Indian National Science Academy (INSA).
2. I borrow the terms "Anglicist" and "Orientalist" from Baber (1996).
3. The Indian National Congress (INC) (now called the Congress Party) is the oldest surviving political organization in India. The party was founded by Allan Octavian Hume in 1895. It played a major role in the nationalist struggle toward independence, and past presidents include luminaries of the independence struggle such as Mahatma Gandhi, Motilal Nehru, Sarojini Naidu, Lala Lajpat Rai, Jawaharlal Nehru, and Dr. Annie Besant. It was elected the ruling party for most of India's post-Independence era, at various times under the leadership of the Nehru-Gandhi dynasty, including Jawaharlal Nehru, his daughter Indira Gandhi, and her son Rajiv Gandhi. Following the 2004 elections, the Congress Party, led by Sonia Gandhi, the widow of Rajiv Gandhi, has returned to power by forming a coalition government in the country.
4. Quoted in Baber (1996, p. 184).
5. Ibid.
6. Thomas Macaulay was a British intellectual who entered Parliament in 1830 and took a job at the Supreme Council of India in 1834. He was instrumental in redrafting the Indian penal code, in addition to instituting English instruction in India.
7. Quoted in Singh (1988, p. xvii).
8. Five-Year plans are the Government of India's bidecennial projected growth strategy. The country is currently into its eleventh Five-Year Plan. The Five-Year Plans were modeled after the Soviet Union's Five-Year Plan system.
9. University of Roorkee is included in this list because 1 year after this report was released, the university, one of the oldest in the nation, was given the status of IIT.

Chapter 3

Transnationalism Among Indian Software Entrepreneurs

New Modes of Immigrant Adaptation

Writing about India in 1917, Weber extended his thesis on the relationship between religion and capitalist initiative by arguing that the centuries-old caste hierarchies in India, sanctioned by Hinduism, would be a barrier to the development of the rational, bureaucratic structures necessary for the establishment of modern capitalism in Indian society. The growth of capitalistic patterns of behavior within "traditional" societies and, by extension, among members of such societies, was far from being an anticipated outcome. Even though he conceded that significant transformations were underway, nonetheless, in the pessimistic tone that often runs through his expositions, he maintained that it would be a long and slow path for changes in the social realm in India to translate into shifts in the economic sphere.

> [I]n the district of Calcutta, old Europe's major gateway, many norms have practically lost their force. The railroads, the taverns, the changing occupational stratification, the concentration of labor through imported industry, colleges, et cetera, have all contributed their part. The "commuters to London," that is, those who studied in Europe and who freely maintained social intercourse with Europeans, used to become outcasts up to the last generation; but more and more this pattern is disappearing. And it has been impossible to introduce caste coaches on the railroads in the fashion of the American railroad cars or waiting rooms which segregate "White" from "Black" in the Southern States. All caste relations have been shaken, and the stratum of intellectuals bred by the English are here, as elsewhere, bearers of a specific nationalism. They will greatly strengthen this slow and irresistible process. For the time being, however, the caste structure still stands quite firmly. (Weber, 1946, p. 397)

A few pages later, he declared: "Modern industrial capitalism, in particular the factory, made its entry into India under the British administration and with direct and strong incentives. But comparatively speaking, how small is the scale and how great the difficulties!" (Weber, 1946, p. 414).

Between the time of Weber's observations and the contemporary period, two things are notably different. First, it is not London, or the cities of Great Britain, that represent an important node in the transnational mobility of educated Indians, those who maintain "social intercourse" with the West. Rather, the primary port of call for educated migrants, including the hyper-mobile transnational Indian entrepreneurs, is the United States and, especially, Silicon Valley.

Secondly, despite Weber's predictions for the "great difficulties" that lay in store for the development of an entrepreneurial class, the global software sector has energized the entrepreneurial capacity of significant numbers of Indians, both as immigrants in the United States, and within the home country. In an era of rapid technological change, the "factory" is no longer the site that signifies economic progress; in the contemporary "informational society," to borrow Castell's term, countries strive to compete in the knowledge industries. Weber might be excused for not being able to project into the future and predict that the trajectory that would lead significant numbers of Indians to entrepreneurial activity would be through the information-technology (IT) sector. It is worthwhile to bear in mind that the majority of entrepreneurs who compose the entrepreneurial class in the software industry do not have roots in the traditional industrial powerhouses of India; instead, for these middle-class immigrants, the impulse for self-employment rests on their education and training as technologists and engineers.

Even if the presence of Indian software entrepreneurs runs contrary to Weber's claims, his idea that *rationality* is the driving force and a fundamental prerequisite for capitalist growth still holds true in the case of the Indian software entrepreneurial class. The preceding chapter outlined how the Indian state, building on colonial policies from a preceding era, not only committed itself to promoting higher education in science and technology—thereby generating a large cadre of highly trained and skilled personnel—but also based such education on the premise of inculcating a scientific, and therefore "rational," temperament among its pupils.

This chapter makes two points. One, that the presence of Silicon Valley's immigrant Indian entrepreneurs should be seen within a larger context of labor migration between the United States and India. The increasing globalization of the software industry has dispersed IT work around many geographical locations. India's emergence as a global provider of IT solutions for companies around the world has had an impact on both immigration patterns between India and the United States as well as the emergence of immigrant entrepreneurs in Silicon Valley. Second, the best way of conceptualizing immigrant entrepreneurship in an era of increasing globalization is through the theoretical lens of transnationalism.

From the Movement of People to the Movement of Jobs: "Brain Drain" to "Bodyshopping" to Outsourcing

The migration pattern between India and the United States has long been termed "brain drain" because of a loss of human capital for a developing country (in the form of skilled workers and students) and a gain in labor personnel for a developed one (Ananth, Babu, & Natarajan, 1989; Bhagwati & Partington, 1976; Kabra, 1977; Khadria, 1999; Sukhatme & Mahadevan, 1987). The emigration of skilled workers continues and the number of students arriving in the United States from India has increased in the last two years, with India now the leading sending country to U.S. colleges and universities.

However, a closer look at immigration patterns in the information-technology industry reveals that there has been a shift in the last 10 years from a movement of people to a movement of jobs.

By the late 1980s, and accelerating in the 1990s, the increasing labor shortages in the information-technology sector in the United States created a need for IT professionals, most of whom arrived from India (Immigration and Naturalization Service [INS], 2001). The flow of workers in this particular sector was given its own term—bodyshopping—to describe both the needs of corporations facing manpower shortages and the nature of work done by Indian immigrants (Dedrick & Kraemer, 1993). A pejorative term, bodyshopping was used to describe the movement of workers (primarily from India) who did little more than provide inexpensive labor on an hourly basis for low value-added programming services at customer sites (in the United States). According to Lakha, in 1990, over 95% of Indian software companies were involved in bodyshopping activities, and of the 3,000 programmers who were working in the software export sector, the majority of them were on assignment abroad (Lakha, 1995).

The term bodyshopping is rarely heard anymore. Rather than having labor come to them, as was traditional in the era of brain drain and bodyshopping, companies are increasingly comfortable with sending jobs overseas, thus quelling the urgency with which Indians immigrated to the United States to seek better job opportunities and increasing the public outcry in the United States over the loss of white-collar jobs.[1] The section that follows describes the shift from brain drain to bodyshopping to outsourcing, and its impact and implications for the migration trends between India and the United States.

Indian immigrants to the United States began to arrive in significant numbers beginning in the late 1960s. This was a time when the tumultuous changes of the Civil Rights Movement addressed not only the situation of people of color within the United States, but also the issue of immigration restrictions that had thus far barred entry from much of Asia and Latin America. Since the mid-1960s, immigration from Asia has renewed on a large scale, and the 1965 immigration reforms have altered the migration landscape—as appendix C shows, the top five sending countries in the period 1981–1997 include China, India, and the Dominican Republic, none of which had significant emigration (compared to countries within Europe) in the decades before the reforms. The only other period that witnessed a similar influx from Asia started with the gold rush in the mid-19th century and extended to 1934[2] (Ong, Bonacich, & Cheng, 1994).

The year 1934 is significant because it marked the culmination of a series of restrictions placed on Asian immigration. The Chinese Exclusion Act of 1882 restricted any further immigration from China, the Gentleman's Agreement of 1907–1908 denied entry to Japanese and Koreans, the 1917 Immigration Act meted out the same treatment to Asian Indians,[3] and finally, the Tydings-McDuffie Act in 1934 added Filipinos to the list of excluded nationalities. A provision in the 1924 Immigration Act reaffirmed these prohibitions by banning the admission of persons ineligible for citizenship, a category that included all Chinese, Japanese, Koreans, and Asian Indians. It was these racist legal restrictions that the 1965 reforms sought to dismantle, creating the conditions for the entry of Indians (among other groups) to the United States (Takaki, 1989).

The census data of 2000 estimates 1.7 million Asian Indians in the United States, composing 0.6% of the total population.[4] Despite the miniscule number, the 1990s have proved transformative. Asian Indians are the fastest growing group, as there has been a 106% increase from the 1990 census (U.S. Census, 2000). Table 7 shows that prior to the 1965 Immigration Reform, there was a very small presence of Indians in the United States,

Table 7. Historical enumeration of immigration from India to the United States.

Years	Number of persons entering the United States
1820–1900	696
1901–1960	12,911
1961–1970	27,189
1971–1980	164,134
1981–1990	250,786
1991–2000	363,060

Source. Immigration and Naturalization Service, 2001.

an aspect that has changed rapidly in the last 4 decades, particularly beginning in the 1990s. Appendix D demonstrates another significant shift—while it used to be the United Kingdom that received the bulk of Indian emigrants, once the United States allowed legal migration, the outflow of Indians steadily changed course, and it was the United States, not the United Kingdom, which became the prime destination.

The Brain Drain From India

The change in the law in 1965 had a specific aim—while it abolished the national origin quota system, thus eliminating national origin, race, or ancestry as a basis for immigration to the United States, the allocation of immigrant visas gave preference to relatives of U.S. citizens (for the reunification of families) and "*for persons with special occupational skills, abilities, or training (needed in the U.S.)*" (Emphasis added) (Immigration and Nationality Act Amendments of October 3, 1965). As a result, unlike previous immigrant flows into the United States that comprised mainly unskilled labor, in the post-1965 phase, it was primarily urban, educated, English-speaking and highly skilled Indians who entered the country, hence the use of the term "brain drain." Jensen stated, "Almost a hundred thousand engineers, physicians, scientists, professors, teachers, and their dependents had entered the U.S. by 1975" (Jensen, 1988, p. 280). Furthermore, this pattern of migration from India was not unique; rather, the flow of professional and technical workers to the United States was an integral part of the larger migration process between the United States and other countries in Asia, notably China, Taiwan, Korea, and Hong Kong (Cheng & Yang, 1998; Ong, Cheng & Evans, 1992).

The data in table 7 reflect the pattern of a steady movement of a highly educated labor force from India to the United States, one that increased in number in the last decade as the software industry gave rise to specific labor needs. The educational attainment, income, and occupational level indices of the Asian Indian population reflect the pattern of selective immigration.

Indians compose one of the wealthiest and best-educated communities in the country, with a median household income of $60,093. Compared with $41,110 for white families,[5] few groups have comparable income levels. While in 1990, the income of Japanese was the highest among any Asian group and exceeded that of non-Hispanic whites, by the year 2000, Asian Indians were out-earning Japanese Americans. Indian American households also have a high proportion of dual- and triple-income homes. In fact, close to 20% of Indian American households have three or more workers. Furthermore, more than 67% of foreign-born Indian Americans hold advanced (postbachelor's) degrees. Among their American-born counterparts aged 25 and older, only 21% hold postbachelor's degree accreditation. Among Indian Americans in the workforce in 1990, 30% were employed in professional specialty occupations, compared with 13% for all U.S. employees (U.S. Census, 2000).

Bodyshopping of Software Professionals

The movement of skilled workers that began in the mid-1960s accelerated in the 1990s as a result of changes in the software industry, a time when hi-tech IT professionals were in short supply in the United States, and Indians arrived to fill an unmet need in greater numbers than from other countries (INS, 2001). Until recently, migration patterns of highly skilled workers has received little attention among international migration scholars. During the 1970s, attention was focused on brain-drain studies, mainly within policy circles. However, increasing economic globalization, the internationalization of firms, and the transnational mobility of labor have created a need for greater research on the issue of migration of high skilled workers (Iredale, 2001; Koser & Salt, 1997).

Research has located the migration of the highly skilled within broader economic processes of global restructuring. Very few studies focus on the decision-making processes of individuals, although there are some exceptions (Findlay, 1995; Li, Findlay, Jowett, & Skeldon, 1996). Far greater attention has been devoted to macrolevel models that analyze the role of transnational corporations in intrafirm mobility of employees (Lawson, 1992), the role of the state in controlling immigration of skilled workers (Cornelius, Martin, Hollifield, & Tsuda, 2003), the internalization of professions and higher education (Iredale, 1999; Phillips & Stahl, 2001), and the role of bilateral and multilateral agreements in facilitating the flow of highly skilled labor (Iredale, 1999).

I examine the mobility of high-skilled labor from India to the United States by placing it within the context of the globalization of the software industry. Specifically, the change from bodyshopping of work to outsourcing can best be understood if consideration is given to the nature of work performed by Indian software houses, and thereby of employees within these corporations. Using Gereffi's value chain (1994) as a starting point, I show that Indian software companies were able to provide higher value-added services, and this, in turn, created the shift from the movement of people (bodyshopping) to a movement of jobs (outsourcing).

Two changes in the computer industry had a role to play in the significant rise in immigration of highly skilled workers from India in the 1990s. Software now accounts for more than 85% of the cost structure of a total computer system in comparison to less than 15% in the late 1950s. This is because software production has remained relatively craftlike and labor intensive (Gibbs, 1994). In sharp contrast to hardware production, which has become highly automated, software development is prone to "bugs," delays, and cost overruns. In other words, software productivity and quality have lagged hardware, creating a "software bottleneck." The industry needed significant manpower to oversee the operations of the software industry.

Secondly, in addition to the increased need for professionals simply to keep up with the growing needs of the software industry, there was also, in the 1990s, the realization that systems needed to be adjusted to accommodate the change in date when the year 2000 arrived, an issue that came to be known as the Y2K problem. This peculiar problem was confronted by large sectors of the U.S. economy and was not simply restricted to the IT industry. As a result, it was estimated that in 1997, 190,000 jobs were waiting to be filled in the United States. (Baker, 1997). Together, these two industry demands created the need within U.S. corporations to engage in the bodyshopping of immigrant IT workers from India.

To understand the negative connotations of the term, it is useful to place the nature of work performed by Indians within the overall software production cycle. Software refers both to the instructions that direct the operation of computer equipment and the information content, or data, that computers manipulate (Schware, 1995). The process of developing a final software product involves a series of steps. Figure 1 presents these steps in a diagrammatic form, with the caveat that software production is more complicated and the actual steps are not so clearly defined.

The most creative aspects in this process, and therefore the stage that is the most "value-added" (Kogut, 1985), comes at the early steps in identifying the problem and the solution, and designing the system. Programming or coding, testing, installation, and maintenance are at the lower end of the value chain (Heeks, 1996, pp. 81–83). When companies faced manpower shortages, they had to look overseas to fulfill critical labor demands. However, reluctant to relinquish creative control, several firms in the United States decided to contract out the relatively unproblematic tasks of coding and testing, and to have the work carried out onsite.

Companies with shortages relied on the H-1B visa program, which allows skilled professionals to be "imported," and it was India that was the largest sending country in this period and under this program (INS, 2001). Such was the demand for skilled workers that in October 1998, through legislative fiat, the cap in the United States was raised from bringing in 65,000 software

FIGURE 1. Stages in software production.

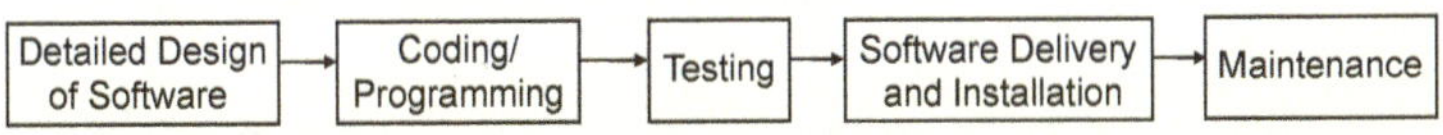

Source. Lateef (1997, p. 4).

specialists each year to 95,000 in 1998 and up to 105,000 every year until 2002 (U.S. Citizenship and Immigration Services, 2003). As Castells (1996) rightly pointed out:

> There is, increasingly, a process of globalization of specialty labor. That is, not only highly skilled labor but labor which becomes in exceptionally high demand around the world and therefore, will not follow the usual rules in terms of immigration law, wages, or working conditions. This is the case for high-level professional labor: top business managers, financial analysts, advanced service consultants, scientists and engineers, computer programmers, biotechnologists and the like. (p. 212)

Indians relied on two visa categories for legal entry—the H-1B and the L-1 visa programs. Understanding the numbers associated with the H-1 B visa category provides an explanation for the quickening pace of Indian immigration in the 1990s. According to the 2001 INS data, the largest proportion of H-1B petition workers approved (see footnote for explanation of the category H-1B worker)[6] was born in India, composing 48% of the total; the second largest proportion was born in China, a mere 9%. Of the total H-1B visas granted each year to workers worldwide, more than a third are for employment in computer-related industries. Within computer-related occupations, 71% of H-1B beneficiaries were born in India. Without India, the percentage of beneficiaries employed in computer-related occupations drops to 32%. In no other occupation did any single country have a majority of beneficiaries. Other characteristics of beneficiaries in computer-related occupations were a median age of 28 and median income of $58,000 (INS, 2001).

In the 1990s, software professionals from India took advantage of the job opportunities available in the United States in computer-related industries, as highlighted in table 8. The table compares the numbers of Indian software professionals admitted to this country with the numbers arriving from China, the second-largest sending country. The startling contrast in numbers is testimony to the fact that the hi-tech industry was especially reliant on the skilled labor force from India.

Furthermore, appendix E lists the top 20 leading employers of H-1B workers between October 1999 and February 2000. The list includes three Indian-owned companies in the United States (Mastech, Syntel, and Tekedge) and four software companies based in India with subsidiaries in the United States (Wipro, Tata Consultancy Services, Infosys, and Tata Infotech[7]), further demonstrating that not only were Indians arriving as employees but that Indian-owned software businesses were involved in the immigration process.

H-1B visas can only be granted when an employer sponsors the legal documents necessary for immigration. Although the visa is meant for a temporary period of stay, many Indians

TABLE 8. Comparing H-1B beneficiaries from India and China (selected years).

	Number of Persons	
Year	**India**	**China**
1992	8,246	2,731
1993	11,411	2,749
1994	16,948	2,721
1995	22,309	3,497
1996	29,239	4,377
1998	62,544	7,746
1999	102,453	14,874
2001	104,543	17,192

Source. Immigration and naturalization service, statistical yearbook (various years).

who arrived under this visa category were absorbed by the American labor market and converted their status to that of permanent resident. In addition to the H-1B visa, Indians also arrive under the L-1 visa category. The L-1 visa category applies to aliens who work for a company with a parent, subsidiary, branch, or affiliate in the United States. These workers come to the United States as intracompany transferees who are coming temporarily to perform services either in a managerial or executive capacity (L-1A) or which entail specialized knowledge (L-1B) for a parent, branch, subsidiary, or affiliate of the same employer that employed the professional abroad.

Although originally targeted toward large U.S. multinational corporations, this is an appropriate method for companies of all sizes to seek immediate immigration benefits for their qualifying employees. Further, the L-1 visa may provide quick access to lawful permanent resident (immigrant) status in the United States. The L-1 visa is more selective, as it can only be granted to professionals who work in an executive or managerial capacity, and compared to the H-1B visa, there are no limits on how many professionals enter under this category. Another important difference with the H-1B visa is that companies who bring workers with an L-1 visa do not need to pay the prevailing American wage to these visa holders. Indian companies that used to rely on the H1-B visa program now increasingly use this other category—it is estimated that 7% of all L-1 holders are of Indian origin. Professionals with L-1 visas are the hypermobile subset of the already mobile software professionals. The L-1 visa lets workers enter the country multiple times over the year (Konrad, 2003).

Thus, both the H-1B and L-1 visa programs enabled companies to rely on a foreign labor pool to control and manage the problem of labor shortages in the United States. However, the pattern of bodyshopping for employees is no longer the preferred route for conducting work in the information-technology sector. Corporations have begun to rely on outsourcing to meet their technology needs, and India is the leading country to which companies outsource (Chopra et al., 2004).

The Maturation of the Indian Software Industry: Outsourcing

Outsourcing can involve different kinds of business arrangements. It may mean that an organization has opened its own subsidiary company in India, so even if work is contracted out, it remains within the parent organization.[8] Another route to outsourcing is to contract work to a third-party company. In some other cases, an outsourcing arrangement may result in a new company being formed that is a joint venture between the supplier and the client[9] (Kobayashi-Hillary, 2004).

What all of these outsourcing situations have in common is that the actual work is conducted in India, or, even if conducted onsite at the customer's location (say, in the United States), the workforce is Indian. Thus, even though jobs are relocating to India, employees in the IT sector in India continue to be highly mobile. Data on the top three Indian IT companies—Tata Consultancy Services (TCS), Wipro, and Infosys—provide a startling picture of the intense mobility patterns of highly skilled workers in India. "nearly a quarter of TCS' 24,000 software professionals are abroad for periods of between two weeks and two years. About half of Wipro's 13,000 IT engineers are overseas on short assignments, while some 4,000 of Infosys' 14,000 engineers are on overseas assignments at any one time" (Merchant, 2003). The short-time assignments are possible because of the availability of immigration categories such as the L-1 visa in

U.S. immigration law.[10] When Indian employees work on-site in the United States, they are paid the wages offered by the Indian company and are not legally bound, under the L-1 visa requirements, to pay prevailing American wages. These numbers also attest to the changing nature of work in knowledge-based industries—rather than directly hire an immigrant Indian worker, companies are increasingly hiring Indian companies, and thereby, the Indian personnel who are employees of that company.

One of the reasons why outsourcing activities have surged in the recent past is the cost differential between hiring a software specialist in the United States and one in India. Table 9 indicates India's cost advantage relative to other competing locations. A second competitive advantage is the 12-hour time difference between India and the West Coast of the United States, thereby allowing technology companies in Silicon Valley to operate on a 24-hour work cycle (Abraham & Manning, 2004; Lateef, 1997). However, cheaper labor and the work cycle are not the only reason why companies are shifting work to overseas locations.

TABLE 9. Comparing India's export industry size and cost per employee with other outsourcing destinations, 2003.

	Export Industry Size (U.S. $ million)	Cost per employee (U.S. $)
India	8,955	5,000–12,000
Canada	3,780	36,000
Ireland	1,920	25,000–35,000
Israel	900	25,000
Philippines	640	7,000
South Africa	96	18,000

Source. NASSCOM IT Industry Fact Sheet, 2003.

Figure 2 is a value chain for the software industry and is a good starting point to analyze how, within a decade, companies in India were able to move from being the suppliers of global manpower to enticing white-collar jobs away from developed nations. Following the work on producer-driven and buyer-driven global commodity chains (Gereffi, 1994), O'Riain delineates the various segments in the technology-driven commodity chain (2004). The most creative stage of the

FIGURE 2. Software-industry value chain.

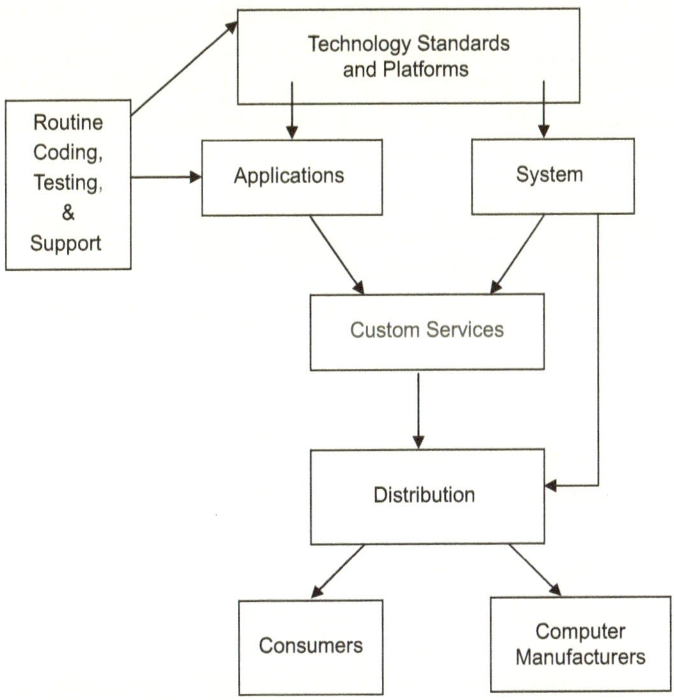

Source. O'Riain (2004).

supply chain is the technology standards and platforms segment. Control over technical standards is where the power rests in the global software value chain. Whether a new technology standard is broadly accepted and employed by various technology companies is key to the survival of new products, and thereby, of the IT companies that develop the products. Maintaining high value within the chain is software applications, which involve the sale of standardized products. Systems integrators use various resources to provide a technology solution to a particular customer, drawing on many other segments of the chain to provide this solution. Custom services companies offer specialized products to fit customers' exact needs, and may even write new code. This segment requires skilled programmers and people who know the existing software very well. Coding, testing, and support of IT products are the part of the process that is at the least innovative end of the chain, even if the most visible (O'Riain, 2004).

In the early stages of the globalization of the software industry, the segment that was most likely to be outsourced was routine coding and testing. Companies in India were regularly given the task of performing these operations. What has changed in the last decade is that many companies in India have moved up the value chain and are able to provide customized services, be reliable system integrators, and, importantly, offer applications sold as standardized products. Indian-owned companies are unable to penetrate the arena of creating technology standards, largely due to the financial constraints involved in marketing (Ram, 2004). Nevertheless, the growth of the software industry in India has been remarkable in the range of services the industry as a whole can now offer. An employee of SAS Institute in Cary, North Carolina, explained the nature of work undertaken by employees of SAS in the company's India location:

> For instance, in my division (Analytical Solutions), several product groups are doing joint development with SAS India in Pune. Work is shared across distances because of the time difference. People at the other end (in India) are also SAS employees. They are not simply taking orders, but they also initiate solutions. Regarding skill sets in India—one of the things we are developing—business problem solving—is raising the level of skills to more than just programming, for instance, developing business domain expertise in areas such as financial risk management, retail management, anti-money laundering, Sarbanes-Oxley corporate governance etc.[11]

Thus, as the software industry in India has matured over the past decade, Indian companies and their employees have risen to become the leading technology-solution providers to companies across the globe. This has led to increased mobility of IT professionals and a growth in the entrepreneurial aspirations of Indians, both as immigrants in the United States and in India.

It is clearly evident from the previous section that India has been a source country for a skilled, technically qualified labor force, and that Indians in the United States have attained levels of professional mobility that few other ethnic groups can compare themselves to. The section that follows is a detailed examination of how, over time, Indians chose yet another mode of achieving economic mobility. The 1990s not only witnessed the influx of software professionals from India to the United States, but it also gave rise to the transnational entrepreneurial Indian. The growth of self-employment within the Indian software community resulted in a different kind of flow—not only one of people, but also a movement that includes goods, knowledge, and information.

THE EMERGENCE OF THE TRANSNATIONAL INDIAN ENTREPRENEUR

The growth of Indian entrepreneurship is best viewed through the lens of the emerging field of transnationalism. Vertovec's (1999) definition of transnationalism is a useful starting point:

> Transnationalism describes a condition in which, despite great distances and notwithstanding the presence of international borders (and all the laws, regulations and national narratives they represent), certain kinds of relationships have been globally intensified and now take place paradoxically in a planet-spanning yet common—however virtual—arena of activity. (p. 447)

In delineating a new arena of study, Portes et al. (1999) caution researchers against some of the pitfalls that can lead to the devaluation and loss of the heuristic value of the concept. It is necessary, therefore, to establish and delimit the phenomena in question. Three conditions are necessary to demarcate the novelty of the process and to justify the need to describe activities as an "emergent research field": (a) the process involves a significant proportion of persons in the relevant universe; (b) the activities of interest are not fleeting or exceptional, but possess certain stability and resilience over time; and (c) the content of these activities is not captured by some preexisting concept, making the invention of a new term redundant.

A comparative examination of two surveys, conducted a decade apart, establishes the phenomenon of the transnational Indian software entrepreneur. Also, the surveys reveal that the growth of entrepreneurship is a new mode of economic activity for immigrant Indians, and that a significant number of these immigrants are involved in business and social practices that can best be captured by the term "transnationalism," and that earlier concepts are inadequate to describe the phenomenon.

In the most comprehensive study on the subject, a survey published in 2002 by the Public Policy Institute of California on foreign-born professionals in the Silicon Valley region found that 60% of Indians had indicated that they were involved in running their own business either full time or part time (see table 10). Indians have almost equal rates of entrepreneurship in Silicon Valley as native-born workers, and higher than those of either Taiwanese or Mainland Chinese professionals, the latter two being the other significant minority groups composing Silicon Valley's workforce.[12]

On the same survey, when asked about their entrepreneurial ambitions, Saxenian found that Indian immigrants were remarkably enthusiastic about engaging in entrepreneurial activities, remarkable because "the survey was administered during one of

TABLE 10. Comparing Indian entrepreneurship rates with Taiwanese, mainland Chinese, and native-born professionals in Silicon Valley, 2001.

	Full-time (%)	Part-time (%)	Never (%)
Indians	43	17	40
Taiwanese	34	17	49
Mainland Chinese	15	17	59
Native-born	48	15	37

Source. Saxenian (2002).
Note. The question posed on the questionnaire was: Have you ever been involved in founding or running a start-up company?

the most unfavorable times for financing start-ups in the recent history of Silicon Valley" (Saxenian, 2002, p. 19). That the inclination toward entrepreneurship is sustained even during a period of economic recession answers Portes' concern that the activities are not "fleeting or exceptional"; to the contrary, the sustained period of engagement in self-employment attests to the resilience of the phenomenon. Indians demonstrated a greater inclination toward entrepreneurship compared with native and other foreign-born respondents (see table 11).

It is clear from the survey results that Indians are an integral part of Silicon Valley's techno-entrepreneurial culture, in addition to being an important part of the workforce as employees in the hi-tech industry of the region. An important link that is not evident from the previous tables, and which makes the numbers even more striking, is that the rise of entrepreneurship among Indian immigrants is a new phenomenon.

Merely the presence of Indians in the information-technology industry is *not*, in and of itself, surprising, given that ever since the migration from India began in the mid-1960s, it has always involved the movement of a highly educated labor force with expertise in technical fields. Silicon Valley would be an obvious

TABLE 11. Entrepreneurial interest among foreign and native hi-tech professionals in Silicon Valley, 2001.

	Yes (%)	Never (%)	Don't Know (%)
Indian	74	2	24
Mainland Chinese	53	9	38
Native-born	48	13	41

Source. Saxenian (2002).
Note. The question posed on the survey was: Do you have plans to start your own business on a full-time basis?

attractive locale of relocation for this group of immigrants. What *is* surprising, however, is that while previously the migrants incorporated themselves as professionals and did not show an inclination toward self-employment, the last decade has witnessed the rise of the entrepreneurial Indian.

Two studies conducted in the 1980s on alumni of the Indian Institute of Technology (IIT), both sponsored by the Department of Science and Technology of the Government of India, give us further insight into the evolving nature of Indian immigrant incorporation. The aim of both studies was to ascertain the extent of brain drain among the best and brightest students from the premier state-run institutes of technology, an issue of understandable concern for Indian policymakers and educators alike. As part of each study, respondents were asked to indicate the type of job they were currently engaged in. Both studies concluded that entrepreneurship was not the preferred choice among the alumni of the institutions, who had emigrated out of the country. As immigrants, the former students of IIT chose professional occupations, and worked as engineers, academics, or as managers (Ananth et al., 1989; Sukhatme & Mahadevan, 1987).

The study conducted in 1989 by Ananth et al. tracked the migrating patterns of the alumni of the Madras campus of the Indian Institute of Technology from 1964 to 1987, the former being the year the institute graduated its first incoming class. The authors' findings reveal that the alumni were primarily engaged as professionals or academics in their broad areas of specialty; only a small percentage ventured into self-employment (see table 12).[13]

A previous study in 1987 produced similar results. The study contacted the alumni of the Indian Institute of Technology at another campus, Bombay, who had graduated in the 5-year period of 1973–1977 with a bachelor's degree in engineering.

TABLE 12. Percentage distribution of the nature of employment of the alumni of the Indian Institution of Technology (Madras) residing in the United States, 1964–1987.

Category	Academic	R&D	Engineer	Manager	Entrepreneur	Other
Group A	21	28	12	17	2	20
Group B	24	22	11	25	3	15

Source. Ananth et al. (1989).
Note. Group A refers to alumni who left India immediately following the completion of their degree, and Group B refers to alumni who worked in India for a period of at least 6 months before immigrating.
The "Other" category includes those engaged in consulting, finance, or planning.

Of the total pool of respondents, 31% had immigrated out of the country. Of those living abroad, 83% had chosen the United States as their second home. One of the study results was that "the dominant professional status is executive/manager and practising engineer" (Sukhatme & Mahadevan, 1987, p. 22). Of the 140 respondents (see table 13), those who had become entrepreneurs were the fewest, behind executive/managers, practising engineers, consultants (employees), research engineer/scientist, and academic (Sukhatme & Mahadevan, 1987).

One might argue that the populations being compared in table 11 with tables 12 and 13 are not strictly compatible, and it might be erroneous to draw conclusions based on these data. It is true that the 2001 survey (table 11) was administered to the foreign-born population working in the hi-tech sector in Silicon Valley, and makes no distinction regarding institutional affiliation of the respondents, whereas tables 12 and 13 are the results of studies conducted on the alumni of one particular institute (IIT), regardless of where they reside in the United States. However, the comparisons may be justified on the grounds that, as indi-

TABLE 13. Occupational preference of the alumni (1973–1977) of the Indian Institute of Technology (Bombay) residing in the United States.

Category	Manager	Engineer	Consultant	Research	Academic	Entrepreneur
Number	41	39	17	13	13	10	7

Source. Sukhatme and Mahadevan (1987), (N = 140).

cated in chapter 2, the alumni of IIT are an elite group, not only within India but also in Silicon Valley. With extensive alumni networks, and a reputation that is now commonplace, currently IIT-ans show high rates of entrepreneurship (see appendix B). Therefore, the fact that alumni of IIT in the 1970s and 1980s were not inclined toward entrepreneurship can be taken to be a good indication that indeed, highly skilled Indian immigrants were not seeking entrepreneurship as an avenue of economic mobility before the 1990s.

Table 14 highlights the rapid growth of Indian-owned hi-tech companies in Silicon Valley in the 1990s.

What the comparison reveals is that there has been a marked change in trajectory of immigrant incorporation for Indians in the hi-tech industry, and increasing numbers view entrepreneurship as a viable economic option. Of course, it is not enough to witness the rise of entrepreneurship to justify the use of the term "transnational." Glick Schiller, Basch, and Szanton-Blanc (1992) offered a definition of transnationalism that limits the use of the term for the activities of migrants:

> We have defined transnationalism as the processes by which immigrants build social fields that link together their country of origin and their country of settlement. Immigrants who build such social fields are designated "transmigrants." Transmigrants develop and

TABLE 14. Indian-run companies as share of total Silicon Valley hi-tech start-ups, 1980–1998.

Year	1980–84		1985–89		1990–94		1995–98	
Start-ups	No.	%	No.	%	No.	%	No.	%
	47	3	90	4	252	7	385	9

Source. Saxenian (1999).

> maintain multiple relations—familial, economic, social, organizational, religious, and political that span borders. Transmigrants take actions, make decisions, and feel concerns, and develop identities within social networks that connect them to two or more societies simultaneously. (Glick Schiller et al., 1992, pp. 1–2)

We need to demonstrate clearly that the entrepreneurs in Silicon Valley indeed conduct their activities in more than one location and maintain multiple relations, cutting across borders. Portes et al. (1999) delimit the concept of transnationalism by referring to the "occupations and activities that require regular and sustained social contacts over time across national border for their implementation" (p. 219).

According to Saxenian, 80% of Indians in Silicon Valley were found to travel to their native country five or more times a year. Thirteen percent of Indians had invested their own money in start-ups or venture funds in India. A further 46% reported helping arrange contracts for businesses in India, and 34% had acted as advisors for companies in India (Saxenian, 2002). The entrepreneurs fit the definition of transmigrants because they live in a social world that creates a seamless tie between their country of origin and country of residence, to the extent that it is difficult to analytically separate the two contexts. And, as scholars (Goldring, 1996; Mahler, 2002; Smith & Guarnizo, 2002) point out, the success of transmigrants does not so much depend on abandoning their culture and language to embrace those of another society as on preserving their original cultural endowment, while adapting instrumentally to a second.

Another recent trend is that not only are increasing numbers of Indians in the hi-tech sectors setting up their own ventures, but they have also begun to express and actualize a desire to return to India or establish networks with their home country. In the same survey conducted in 2001 by Saxenian, 53% of respondents below the age of 35 expressed an interest in returning to India, and 35% and 28% indicated the same in the 36–50 and 50 and over age ranges, respectively. Further, these numbers remained the same when visa status (temporary worker, permanent resident, or citizen) was factored in.

Compared to the numbers indicated in the previous text, it is interesting to note that in the study conducted in 1987, only 24% of alumni surveyed said that they had any intention of returning to India (Sukhatme & Mahadevan, 1987). Although the study did not disaggregate by age, considering that the study was on alumni who had graduated between 10 and 15 years before the study was conducted, the age group of respondents is likely to be under 35 years of age. Thus, more Indian hi-tech workers aged 35 years or less now say that they are likely to return to India, as compared to the situation a decade and a half ago.

Once again, comparing the two studies, separated by a decade, some interesting patterns emerge. As table 8 shows, there is no change over time in stating that there are family reasons, cultural/lifestyle issues, and a sense of national pride/contributing to the home country as being important factors in the decision to return to the native country. A crucial difference that is especially notable is that perceptions about job opportunities in the United States have changed over time. While none of the respondents in the 1989 study indicated that a reason to return to India was because of poor job opportunities in the United States, 28% of respondents in 2001 answered that limits on professional advancement in the United States constituted an important reason in their eventual return to India.

Also, considerably more numbers of immigrant Indians consider that professional opportunities for growth are good enough in India for them to consider giving up their lives in the United States. Brain drain studies consistently indicated that one critical reason for migrating out of India after obtaining an engineering degree from the most reputed technology institute in the country was because of poor job opportunities in the home country (Ananth et al., 1989; Bhagwati & Partington, 1976; Kabra, 1977; Sukhatme & Mahadevan, 1987). The most recent survey shows a remarkable shift in attitudes. Respondents in the 2001 study say that an important reason for returning to the country of origin is the presence of opportunities in the home country. The lack of opportunities in India was why earlier cohorts of skilled Indians were emigrating. Clearly, this has significant implications for India's economic development.

While there are no exact numbers on how many Indians have returned to their country of origin, by one estimate, there are about 35,000 "returned nonresident Indians"[14] in Bangalore, India's first hi-tech center, with many more scattered across India. Many return for the reasons stated in the previous text, to assist in India's economic development and to reconnect offspring with the country and the culture the parents left behind.

Srikanth worked for 14 years in Silicon Valley for a number of different companies, including Sun Microsystems and for his own start-up. He returned to India to form a technology nonprofit, which describes itself on its Web site as, "a new breed of IT Social

TABLE 15. Reasons stated by Indian immigrants in the United States for returning to their home country: Comparing the 1989 and 2001 surveys.

Reasons for Leaving the United States to Return to India			
1989 Study		**2001 Study**	
National Pride	50%	Culture and lifestyle	75%
Lack of sense of belonging abroad	35%	Professional opportunities in India	68%
Family reasons	29%	Contribution to home country	66%
Jobs in India	29%	Family reasons	60%
Cultural reasons	29%	Favorable government treatment	49%
Poor job opportunities abroad	0%	Limits on professional advancement abroad	28%

Source. Ananth et al. (1989); Saxenian (2001).

Startup (a registered not-for-profit trust) whose mission is to provide an eGovernance Software System for use in city municipalities all across India–for free" (http://www.egovernments.org/index.htm). The software platform that Srikanth is developing uses digital mapping to permit accurate property tax assessments and collection. Its uses will allow electronic tax payment, birth and death registrations, the filing of citizen grievances, and the public tracking of small infrastructure projects. Speaking of the current state of India's software technology sector and the spirit that moved him to return, Srikanth explained, "We [India's software professionals] are making a couple of billion of dollars of software exports, but we are not solving India's problems. We are solving the world's problems." An estimated 110 companies were started in Bangalore by these "returnees" by the end of March 2004 (Waldman, 2004a).

The reverse trend is a result, in part, of the outsourcing of jobs from the United States to different parts of the world, and increasingly, technology companies view India as the destination of choice when sending work overseas. Many Indians acculturated themselves to patterns of work and management styles in the United States and are now returning with their accumulated wealth and knowledge to run businesses that still keep alive their connection to the United States. S. Nagarajan returned to India in 2002 and established a call center in Bangalore with 20 employees. In the past 2 years, the number of employees has risen to 3,600 and the annual revenue is $30 million (Waldman, 2004a). What makes entrepreneurs like Nagarajan transmigrants, not simply return migrants, is the fact that the social space they exist in continues to depend on ties with the United States. The companies they run in India carry out work for projects that have been outsourced from the United States and they maintain U.S. citizenship to travel back and forth for business purposes.

I have suggested here that there is a significant shift in the patterns of immigrant incorporation among Indians entering the United States as part of the hi-tech workforce. Indian software professionals have been the single most important source to meet labor shortages in the United States, and both the H-1B and L-1 visa categories are deployed to create increased mobility of these highly skilled workers. Entrepreneurship is increasingly viewed as a means to secure a foothold in the knowledge industry. Increasing numbers of Indians perceive themselves as effortlessly straddling two countries in two different continents, as demonstrated by their economic activities, frequent travels, and, sometimes, eventual return.

Skeptics of transnationalism might argue that there is nothing new in the back-and-forth movement that links countries of destination and countries of origin (Foner, 1997)—such was the normal pattern for European immigrants at the turn of the century, as trans-Atlantic travel was common. However, firstly, historical precedent does not dilute the value of the existing phenomenon. Quoting Merton (1987), "what is more common is that an idea is formulated definitely enough and emphatically enough that it cannot be ignored by contemporaries, and it then becomes easy to find anticipations of it (p. 16)." Secondly, the mobility of the Indian transnational entrepreneur is indeed new and does not have historical precedent. While it would be common for immigrants to send remittances, it was uncommon to find the intense movement that characterizes the contemporary period.

Confronted with the situation in which immigrants are no longer "pushed" out of a country only to be "pulled" into another, the term that best applies to the Indian immigrant software entrepreneur is "transmigrant." Thus, when Portes et al. (1999) argue that "nothing is gained by calling immigrants 'transmigrants,' when the earlier and more familiar term is perfectly adequate to describe the subjects in question" (p. 219), they are exhorting researchers to justify the use of the new term. The push-pull model of immigration is insufficient to explain the experiences of Indians in the software sector. Concepts such as assimilation and acculturation have been extensively used in the sociological literature to enable us to understand the processes through which immigrants adapt to the new surroundings they find themselves in (Alba & Nee, 1997; Portes & Rumbaut, 1996). Furthermore, studies on immigrant entrepreneurship have focused specifically on self-employment to understand processes of incorporation into host society by accounting for variation in entrepreneurship rates among different ethnic groups (Barret et al., 1996; Light & Bonacich, 1988; Light & Gold, 2000; Min & Bozorgmehr, 2000; Waldinger, 2001; Texeira, 2001). Even though the ideas of assimilation, acculturation, or incorporation give us insight into the processes of adaptation, in the case of the highly skilled immigrant in an increasingly networked world, the concepts lack an essential character of individuals whose lives are determined by their incorporation/assimilation/acculturation in *more than one* national arena. Thus, rather than using the term immigrant, or even simply migrant, it is more useful to employ the term transnational because it connotes intense movement and mobility.

The theoretical and practical implications that emerge from the study of Indian entrepreneurs as highly mobile transnationals are the following:

> 1) the emergence of these activities is tied to the logic of capitalist development 2) while following well-established principles of social network development, transnational communities represent a phenomenon at variance with conventional expectations of immigrant assimilation 3) because transnational enterprise is fuelled by the dynamics of capitalism, it has greater potential as a form of individual and group resistance to dominant structures than alternative strategies (Portes, 1999, p. 227).

The emergence of transnational activities of entrepreneurs is indeed tied to the logic of capitalist development—in this case, the organization of the global software industry. The organization of work in this knowledge-intensive industry has generated a highly mobile group of professionals, and concomitantly, allowed the rise of entrepreneurship among Indian software professionals. The shift from bodyshopping to outsourcing in the software industry has quickened the pace of entrepreneurial activities as Indian immigrants find greater opportunities to engage in transnational business practices. Moreover, the globally "stretched" patterns of behavior of Indian entrepreneurs suggests that Indian immigrant entrepreneurs in Silicon Valley do not adopt conventional assimilation methods; instead, their actions allow them to simultaneously coinhabit, both mentally and physically, two different locations.

ENDNOTES

1. Outsourcing became a crucial election-year issue in 2004. Democratic candidate for president, John Kerry, accused the Bush administration of wanting to "export more jobs overseas"; speaker of the house, Dennis Hastert, claimed that "outsourcing can be a major problem for the American worker and the American economy." (Drezner, 2004, p. 1).
2. "Asia" does not include India, or other countries that form part of South Asia. While there were pockets of settlements of Indians in California and Oregon before entry was barred for Indians in 1917, the numbers were significantly fewer compared with immigrants from China, Japan, and the Philippines (Takaki, 1989).
3. In 1946, Asian Indians were granted token immigration quota of 100, but it was not until 1965 that the national origin system (Johnson Act) of 1921 was completely repealed.
4. Currently, there are an estimated 12 million Asian and Pacific Islander Americans in the United States, or about 4% of the total population (compared with African Americans, at 12%, and Latinos, at 11%) (U.S. Census, 2000).
5. "White" refers to non-Hispanic whites.
6. An H-1B temporary worker is an alien admitted to the United States to perform services in "specialty occupations," based on professional education, skills, and/or equivalent experience. A specialty occupation requires theoretical and practical application of a body of specialized knowledge, along with at least a bachelor's degree or its equivalent. For example, architecture, engineering, mathematics, physical sciences, social sciences, medicine and health, education, business specialties, accounting, law, theology, and the arts are specialty occupations. The INS is responsible for the processing and approval of H-1B petitions submitted by domestic employers and their representatives on behalf of nonimmigrant workers seeking temporary employment within the United States. The H-1B nonimmigrant classification permits foreign professionals to enter the United States on a temporary basis to work in their

field of expertise. Under the H-1B program, specialty workers are permitted to be employed for as long as 3 years initially, with extensions not exceeding 3 years. The maximum stay is 6 years (U.S. Citizenship and Immigration Services, 2003).

7. Even though software companies based in India were sending their employees to work on projects in different locations around the world, these same companies are now also attracting talent from overseas. Infosys, one of the top three software houses in India, with 25,000 employees, was established in 1981, and went public in 1993. In 2004, it posted revenues of $1 billion. It now runs a global summer internship program, and last year, it received 8,500 applicants from around the world to fill 75 slots (Waldman, 2004b).
8. SAS Institute, headquartered in Cary, NC, is one such company. SAS has an overseas development office in Pune, India. Employees in Cary and Pune oftentimes jointly work on the same project.
9. A strategic joint venture was formed in 2000 between Satyam Computers, based in Hyderabad, India, and U.S.-based TRW to provide IT services to the automotive industry. Satyam had a 76% share, and the remaining 24% was with TRW, with Satyam managing the operations. In 2003, Satyam bought TRW's shares (Vashistha & Vashistha, 2004).
10. Other OECD countries have similar entry mechanisms for highly skilled immigrant personnel (see Khadria, 2004a).
11. Personal interview with S., SAS employee in Cary, North Carolina, September 2004.
12. The survey was administered online between May–July 2001. The survey sample includes 788 respondents from Greater China, 769 respondents from India, 260 U.S. born, and 189 Taiwanese. These numbers are 3.8, 4.2, 0.1, and 0.3% of the total estimated professionals in the region from Mainland China, India, Taiwan, and the United States, respectively.
13. A total of 5,942 students graduated from IIT Madras during the period 1964–1987. The study obtained information, for comparative purposes, both from students who had left the country and those who were still residing in India. However, as a result of the efforts of the IIT Madras Alumni Association of North America (IITMAANA),

more responses were received from alumni living overseas than those living in India. Altogether, 1,431 correct addresses were obtained—384 in India, and 1,047 abroad. A total of 184 alumni in India and 445 from the United States responded to the questionnaire.

14. They are nonresident Indians because many of them are citizens of the United States.

Chapter 4

Global Ethnography

Methods and Sites of Research

The following chapter lays out the research design for the study on the participation of Indian immigrant entrepreneurs in Silicon Valley. The discussion on selection of field sites begins with the dilemma inherent in conducting ethnographic research on global issues. Given that ethnography has traditionally lent itself to Geertz's "thick description"[1] (1973) of specific sites, how does one understand processes such as transnationalism that, by their very nature, shatter the notion of a single place? Ethnography has had to keep up with the changing social world we inhabit. The idea of conducting "global ethnography" presents challenges and simultaneously allows fresh insight into new social processes at work.

The second and third sections of the chapter deal with the other two axes of conducting ethnographic research—once the field sites are determined, who does one gather information from, and by what means? Therefore, the rest of the chapter addresses the specific details of the research design: the study population, selection of participants, the utility of semistructured interviews and observations, and a discussion on interpretation and analysis of findings.

FIELD SITES: FINDING THE "LOCAL" IN A "GLOBAL" WORLD

The study was an ethnographic inquiry at two sites: San Jose in Northern California and Bangalore in Southern India. It is important to state at the outset that the choice of sites was not dictated by conducting comparative research to ascertain similarities or differences between the two locations. Rather, the research question was guided by an interest in understanding the nature of Indian immigrant participation in the economy of Silicon Valley. However, transnational Indian entrepreneurs are hypermobile, and their activities make them global agents.[2] As indicated in the previous chapter, immigration (albeit in certain contexts, for certain groups) is no longer a one-sided flow between "origin" and "destination" countries, and technically skilled Indian immigrant entrepreneurs have moved beyond the phase of "brain drain" labor movement. The transnationality of the subject population compelled me to become a "transnational ethnographer" and adopt a two-sited field research agenda. Consider the following statement by an Indian entrepreneur in Silicon Valley:

> I don't have a home anymore, or you could say, I have two homes. I split my time between India and the Valley. I have an office in the Valley and one in Bangalore—the one in Bangalore is just being set up, so I have to go there constantly. I am a citizen of the US and am waiting for the Indian government to change its laws to allow for dual citizenship. I make contributions to the temple here and I make contributions to the BJP[3] in India…My son is in boarding school in India.[4]

This statement reflects a constantly shifting sense of "place"—the transnational immigrants in the study are simultaneously members of two different geographical locations on two separate continents. The economic activity of the Indian immigrant entrepreneur—involving one office in Silicon Valley, California, and the other in Bangalore, India—creates a seamless "space," challenging traditional notions of "work*place*." The political identity of the interviewee—United States citizenship with aspirations for dual citizenship in India—further cements the notion of a borderless world and makes unclear which "place" has primacy over the other. The split family setup adds to the duality of the transnational migrant's existence. Even though the example I have chosen is one in which transnationalism is expressed along several different dimensions all at once—economic, political, social, and religious—most of the interviewees in the research study were similarly transnational, at least insofar as their economic activities were concerned.

Thus, conducting ethnographic research on transnational Indian immigrant entrepreneurs entails its own methodological challenges. Gille and O'Riain (2002) described ethnography in the following terms: "[E]thnography explicitly seeks to analyze the social by locating the researcher in the space of the social relations being analyzed." Moreover, they continued, "and this ability to straightforwardly access the social by going to the local becomes problematic under conditions of globalization" (p. 273).

When actions are undertaken in "spaces" of globalization, how does one conceptualize "place"? In other words, since fieldwork research involves specific locations, what is the "local" in a global world? How does one conduct "global ethnography"? Burawoy et al. (2000) and Gille and O'Riain (2002) suggest that sociologists have recently developed a variety of ways of conducting global ethnographies, depending on the lens through which they approach the process of globalization. These studies can be demarcated as those that focus on understanding *global forces, global connections,* or *global imaginations*.[5] Analytically, what distinguishes the three is "whether people experienced globalization as an external force to be resisted or accommodated, whether people participated in the creation and reproduction of connections that stretched across the world, or whether people mobilized and/or contested imaginations that were of global dimensions" (Burawoy et al., 2000, p. 5).

Using Burawoy's typology, the research on transnational Indian entrepreneurs falls under the study of "global connections," where the subject population's economic and social activities are straddling two places at the same time.

Studies that focus on *global forces* examine the ways in which large-scale processes—be they modernity, capitalism, or science—have an impact on localities or communities (Dudley, 1994; Fernandez Kelly, 1983; Gowan, 2000; Scheper-Hughes, 1992). The thrust of the research is to show how the global penetrates the local and how the local responds. In so doing, such studies reveal the mechanisms through which global forces operate. Thus, for instance, the increasing feminization of remunerated work is seen as a product of the growing integration of Third World production systems into the global capitalist network. With the off-shoring of manufacturing work to low-wage countries, a disproportionately large number of female workers have been mobilized in poorer countries (Lim, 1980; Nash & Kelly, 1983; Safa, 1995; Sassen, 1998). These authors document how women (local actors) either assimilate, or resist, the intrusion of overarching processes (capitalism). Detailed ethnographic research has revealed a dual dynamic at work, sometimes contradictory—on the one hand, the women become an invisible, exploited, and disempowered workforce, but their access to wages and salaries (even if low) allows them to gain greater personal autonomy and avoid patriarchal relations in the household (Grasmuck & Pessar, 1991; Prieto, 1992; Wolf, 1992). What is common to these studies is that they explore how particular groups or localities are inextricably tied to events and forces that are beyond their control. The actor, in other words, is at the mercy of external events and processes. The point about the agency of actors is one I shall return to repeatedly in this chapter.

The study of *global imaginations* tend to cluster around how groups—social movements or elites—develop discourses, create visions, and construct identities on what globalization might, or ought to, look like. The actors in studies of global imaginations are more often than not actively engaged in the battle of competing visions of the global (Gille, 2001; Moore, 2001). For instance, Goldman (2001) shows how the World Bank responded to the strong environmental challenges posed to its development projects. The funding for the Bank-sponsored construction of the Narmada Dam in India was withdrawn in the face of stiff anti-Bank protests over the detrimental social and environmental consequences of the dam project. In the aftermath of the Narmada debacle, the Bank developed a new discourse on "environmentally sustainable development." The Narmada movement became a global social movement projecting alternate trajectories of social and economic development, thereby competing with the vision of powerful organizations such as the World Bank. In the ensuing battle, the Bank changed its policies and co-opted its critics' visions by enforcing reform measures within its own organization. Such reform measures were also a result of criticisms emanating from within the upper echelons of the Bank (Stiglitz, 2003). Goldman's study in Laos follows the ways in which the "new" vision of the World Bank was being adopted. In the end, he argues, the Bank merely created a new language but continued to operate in ways that suppressed dissent—experts from nongovernmental organizations and natives were given a voice, but not enough time to carry out the appropriate research. Or, when the knowledge that these experts produced was contradictory to the Bank's goals, it was simply filed away. Thus, studies of *global imaginations* are useful in delineating how, in an increasingly global world, even local knowledge and ideas are shaped by discourses of global organizations.

The third—and most pertinent to the study of transnational Indians—is an examination of *global connections*. Such studies are "from the perspective of those in the middle class from the semi-periphery who have the skills, contacts, and other resources to allow them to incorporate into global economic and discursive flows" (Burawoy et al., 2000, p. 141). What typifies the groups engaged in *global connections* is their involvement in sectors of the global economy that are still emerging, where rules, regulations, and laws have not yet been fully institutionalized. The information-technology sector is an exemplar of one such sector. The groups described as most likely to be studied under global connections—migrants, traders, and technical and occupational communities—fit neatly with the transnational Indian entrepreneurs that form the focus of this book. The perspective used in the study is precisely from the perspective of migrant, middle-class individuals who have the skills in emerging sectors of the world economy, specifically information-technology and the software sector. Through use of skills that allowed them entry into the United States and the mobilization of resources attained as immigrants, Indians have been successful in incorporating themselves into global networks and flows, and, in fact, have been eager participants in shaping global power structures.

It is important to remember, as pointed out in chapter 3, that Indian immigrants arriving under occupational quotas have always had the skills that were valued in the U.S. economy. What makes the present different from the past is that these skills can now be exploited at a global scale. Indian immigrant entrepreneurship in Silicon Valley emerged at the same time that India became a global software center. When software professionals were in high demand in the United States, from the late 1980s and through the 1990s, it was India that was the primary sending country of technically qualified professionals (see table 7 in chapter 3). As more and more companies began to engage in off-shore production, subsidiaries of multinational corporations were established in India (Patibandla & Petersen, 2002). As entire processes of business activities began to be outsourced, India became the leading outsourcing destination (Chopra et al., 2004). At each stage, the Indian immigrant population has had a role to play. India's emergence as the critical node in the increasing outsourcing of business processes has been assisted by the presence of Indian entrepreneurs, venture capitalists, and professionals in the diaspora, primarily the United States, but also Canada and the United Kingdom, according to research sponsored by the World

Bank (Aggarwal, 2004). Indeed, international business scholars are urged to consider topics that are not in the mainstream of the literature on foreign-direct investment (FDI) and its role on developing societies—in particular, the impact of diaspora populations on homeland FDI (Ramamurti, 2004). Thus, middle-class immigrants from India, a country from the "semi-periphery," now find themselves at the center of global connections between their native and adopted countries.

What makes the studies of global connections different from those of global forces and global imaginations is the emphasis on the agency of the actor. Typically, writers on transnational connections emphasize actor-oriented theories and position themselves against the overly deterministic analyses of globalization theorists (Smith, 2001). Writing about urbanization theory in a transnational world, Smith argued that

> the accumulation strategies of capitalist logics, structures, and actors, to which many urban analysts devote so much attention, are not the sole, or at times even the most important, agencies in the constitution of urban life. As important, if not more so, has been the impact of ordinary women and men—their consciousness, their intentionality, everyday practices, and collective action—on the social construction of urban life. (p. 6)

The starting point of the analysis of global connections is usually strategic action, or strategic behavior, expressed by the group. The ethnographer's role, then, is to analyze how these actions/behaviors allow connections to be made across social and political borders.

In designing the appropriate research methodology, the issue of the agency of actors was an important one, an aspect I shall return to in discussing the employment of semistructured interviews. Returning to the issue of field sites, Burawoy et al. pointed out: "While all of us [those who study globalization] are engaged in multi-sited ethnography, while none of us is riveted to a single place and time, it is only in the study of *global connections* that multi-sitedness becomes an object of theorization" (Burawoy et al. 2000, p. 30). If the attempt of ethnographers is to engage themselves in the lives of their subject population, studies of *global forces* can concentrate on a single site. Thus, Blum (2000) can study the effects of globalization on San Francisco's shipyard workers by conducting research at the shipyard itself. The workers at the shipyard, although deeply affected by events in distant locations, are themselves not physically in two places at the same time. In keeping with the lives of their study population, the researcher need not travel to another location for research. The same is true for studies around *global imaginations*, as the focus is on the discourse of globalization, and it is adequate to conduct ethnography at one site where the contestations are being envisioned and acted upon. In the case of the study of *global connections*, however, the ethnographer can only place herself in the lives of the subject population by becoming a transnational researcher herself. However, the problem with the multisited approach is that "from any one site connections fan out in multiple directions, so that the relations between any two sites are usually thin" (Burawoy et al., 2001, p. 30).

To understand the business practices of Indian immigrants, it was not enough to study them only within the context of Silicon Valley. Given the dense network of relationships that they have with the software sector in India, it was also necessary to explore how these connections are being enacted transnationally. However, from Silicon Valley, the connections "fan out in multiple directions"; in other words, Indians in Silicon Valley have business partnerships, social networks, and personal ties that take them to any number of places in India. The following examples of three Indian-owned companies in Silicon Valley highlight the difficulty in pinning down a one-to-one correspondence between any two specific field sites.

Tibco Software is a leading software services company founded in 1997 by an Indian immigrant entrepreneur. It is headquartered in Palo Alto, has 20 offices in the United States, operates in 15 countries in Europe, and is located in 8 countries in the Asia/Pacific Rim. The company has a total of 1,500 employees. The offices in India play a crucial role, as these are "development centers," where Indian software professionals perform critical tasks for Tibco's operations worldwide. In India, Tibco has two locations—one each in the cities of Mumbai and Pune. Take the second example of another Silicon Valley-based company, called eGain. A provider of eService software for the internet, the company was founded in the late 1990s by two Indian immigrants. Like Tibco, the company simultaneously operates in several worldwide locations. In addition to its seven offices in Europe, there is a development center in the western city of Pune, India. Finally, Magma is also an Indian-immigrant-owned venture started in the heart of Silicon Valley, with global operations and a total of 500 employees. Its India operations are concentrated in the southern Indian city of Bangalore.

All three ventures indicated in the previous text share two common characteristics: Their founders are Indian immigrant entrepreneurs based in Silicon Valley, and the companies' operations in India are a critical component of their continued growth and success. Thus, the selection of Silicon Valley as a research site to study Indian immigrant participation in the U.S. software sector was unproblematic. Silicon Valley is not only the center for the U.S. information-technology industry as a whole, but this area in Northern California also has a large concentration of Indian immigrants, many of whom are entrepreneurs (Dossani, 2002; Saxenian, 2002).

However, finding a "local" site in India to study "global connections" was not amenable to any easy solutions. The example in the previous text highlights the varied directions in which these connections are being established. The three companies have established their India offices in various cities—Bangalore, Mumbai and Pune. Other cities in India are equal competitors—Hyderabad, Chennai, and Gurgaon (a suburb of the capital city, Delhi) (Kobayashi-Hillary, 2004).

In the end, I chose Bangalore[6] as my local site in India for several reasons. Bangalore has long been the center of India's information-technology industry, and it is around this city that the first cluster within the country emerged (for how Bangalore emerged as a global software center, see Parthasarathy 2003).

More than 1,000 firms and over half of the world's Fortune 500 companies have an office in Bangalore. Over 50% of the world's SEI CMM Level 5 companies[7] are based there. Major international firms such as General Motors, GE, Dell, and HSBC have facilities, and the headquarters of the top Indian IT companies, Infosys, and Wipro, among others, are in Bangalore. Even though other clusters have emerged, the city continues to be dominant in the information-technology field. In a joint survey conducted in 2004 by KPMG, an international consulting firm and NASSCOM,[8] this cluster in southern India is ranked as the most attractive location for BPO (business process outsourcing) activities, ahead of 12 other information-technology clusters in India. Bangalore was also ranked the fourth most important global hub of technological innovation in the world by the United Nations (Kobayashi-Hillary, 2004). Finally, Bangalore is the preferred destination for those Indians in Silicon

Valley who show an inclination to return to their home country (Dossani, 2002).

Thus, although the interviewees in Silicon Valley did not always have global connections with Bangalore specifically, nevertheless, the decision to go to India to do research was animated by a desire to locate myself in the "space of the social relations being analyzed" (Gille & O'Riain, 2002, p. 273). Altogether, I spent 8 months in Silicon Valley. The first phase was between September 1999 and January 2000, and the second between May and July of 2001. In India, I spent 6 months conducting research in Bangalore, between February and June 2000.

SUBJECT POPULATION AND RESEARCH SETTING

In Silicon Valley, the interviews were with Indian entrepreneurs, top executives in hi-tech companies, managers, and software professionals. The interviews were conducted face-to-face. Some follow-up interviews were conducted over the telephone and over e-mail. Most of these formal office interviews lasted for about an hour. Whenever possible, during the course of the office visit, I would also speak with other executives and the employees of the company.

I also made several visits to two Indian associations in the Valley, specifically TiE, which is the Indus Entrepreneurs, and TANA, Telugu Association of North America, Telugu referring to one of the regional languages of India. These visits would either involve going to their local offices or attending the meetings, conferences, or cultural events organized by the associations.

TiE was started in 1992 in Silicon Valley as a not-for-profit networking organization for entrepreneurs of South Asian origin, but TiE also stands for Talent, Ideas, and Enterprise. The geographical presence of TiE has grown significantly in the last decade with the concomitant rise of the importance of the Indian software industry and the role played by Indian immigrant entrepreneurs in the United States. In keeping with the transnational nature of the industry, the organization now operates in nine countries and has over 40 chapters, including 24 in the United States and 13 in India.

Interestingly, the philosophy of the organization, as stated on their Web site, is that "the values and beliefs of TiE reflect the culture and value system of the Silicon Valley: entrepreneurship and wealth creation are invaluable human endeavors, and successful entrepreneurs find fulfillment in helping budding entrepreneurs" (TiE Web site). As pointed out in chapter 1, Indians in Silicon Valley view themselves to be no different from the "native" population, as their interests, vision, and "culture" are very much in keeping with mainstream society. In this way, Indian immigrant entrepreneurs defy any designation as being "ethnic" entrepreneurs, and this sentiment is reflected in the goals laid out for this important networking organization. TiE has monthly meetings that are well attended; in addition, they hold an annual conference that attracts entrepreneurs from across the country.

From an ethnographer's point of view, attending the meetings, conferences, and events at TiE were an important way to gain insight into the business practices of Indian immigrant entrepreneurs. My role in the meetings had a dual purpose—the first was to conduct observations and the second was to develop contacts. Many of the informal discussions that ensued at TiE functions provided me with the opportunity to delve deeper into not merely the business practices, but also the motivations behind them.

TANA, on the other hand, is a cultural organization for a specific group of Indo-Americans, those who hail from the state of Andhra Pradesh in southern India. Founded in 1977, the objective of the organization is to preserve and propagate the literary, educational, and cultural heritage of Telugu-speaking Indians in the United States. There are several different reasons why I chose this particular organization as a site for ethnographic research on Indian immigrants in Silicon Valley. First, TANA is a good contrast to TiE. While the latter can be considered an important site to understand the economic activities of immigrants, the informal atmosphere at TANA allowed me to study more closely the ways in which Indians in the United States express their social, cultural, and religious aspects of their lives. Secondly, TANA's activities were geared toward the entire family, and this gave me a chance to understand familial and gender practices. Lastly, even though there is little formal data to back this claim, informal evidence suggests that Telugus form a large group of software professionals in Silicon Valley.

In Bangalore, I contacted entrepreneurs and also a subset of return migrants to understand how and why the return journey was made. I attended conferences organized by NASSCOM (the National Association of Software and Services Companies) of India with the same intention as that of attending TiE and TANA meetings—to generate contacts for future interviews and learn more about the research setting through observational techniques. I also had the opportunity to attend two visa sessions organized by the German embassy. These were informational sessions for software professionals on the new immigration regulations devised by the German government to attract Indian software professionals to work in Germany.

ETHNOGRAPHIC DATA: OBSERVATIONS AND SEMISTRUCTURED INTERVIEWS

The ethnographic approach (Hammersley & Atkinson, 1983; Prus, 1996; Stanley, 1990) to social inquiry is interpretive, addressing the nature of social phenomena rather than testing strict hypotheses. Ethnography, according to Geertz (1973), is "not an experimental science in search of laws but an interpretive approach in search of meaning" (p. 5).

Observation allows the ethnographer to understand the research setting. However, observation in and of itself is not an adequate base to performing ethnographic inquiries. It runs the danger of the researcher making "excessive inference and failing to uncover the cognitive processes that produced observed behaviors" (Albers, 2003, p. 22). Coupled with semistructured interviews, the observational notes that I took played a critical role. I was an outsider in all the conferences, meetings, and cultural events I attended. It was not difficult to maintain distance at the business meetings—for the most part, my presence was eyed with cursory interest or bemusement. TiE meetings are an occasion to socialize, but more importantly, they are an opportunity for Indian entrepreneurs or budding entrepreneurs to network with associates in the business world. My presence as an academic was not viewed as a threat or with suspicion. I was welcomed into these meetings and found the occasion to conduct informal conversations, several of which were followed up with formal interviews.

At times, I was asked to involve myself in the cultural events organized by TANA, such as organizing the celebration of Indian festivals or participating in cultural dance shows, but I consciously avoided doing so. I was determined not to engage in participant observation and was content in using these events as a means to engage in minimal, nondisruptive observations from the sidelines. The idea was to familiarize myself with the features of the population and to stimulate questions in my own mind in order to carefully pursue those in future interviews (Adler & Adler, 1994; Spradley, 1980).

Participant observation is one of the pillars of ethnographic inquiry; therefore, a word on why I chose not to adopt this technique is called for. Even though participant observation is more interactive, the consequences of becoming intimately familiar with future interviewees was not a line I wanted to cross. The presence of the researcher can change the behavior of the very people the researcher is trying to observe. Moreover, participant observation is more practical and advisable when the subject population may be wary of the researcher and their intentions, and is suitable when the researcher is trying to gain trust and rapport with interviewees. Given that my research was not on deviant, exotic, or illegal behavior and practices, I did not face the same concerns as those shared by researchers whose subject matter may create grounds for suspicion among interviewees. Participant observation is also a means to bridge the intersubjective gaps that may exist between the researcher and the study population. I had the opposite problem—I was already familiar with the rituals, symbols, and "insider" language associated with the cultural events I was witnessing. What I needed was to distance myself from these in order to view them with the eyes of an ethnographer. I would usually sit as a member of the audience and observe the goings-on. Thus, I used these sites as a way to generate contacts and to take observational notes that were later compiled into research questions to be deployed in the semistructured interviews.

In both Silicon Valley and Bangalore, I conducted semistructured interviews, a total of 42 in the first phase of research at Silicon Valley, 16 in the second phase, and altogether 28 in Bangalore (see appendix F for summary statistics). I had a basic set of questions that asked about educational, professional, and family background, and immigrant history, and questions related to entrepreneurship—why they were in Silicon Valley, how the companies were started, how funding was obtained, what networks were used, what ties the company had to India, if any.

Thus, while there was a common set of questions addressed to every interviewee, the entire interview process would often cover topics and subjects not specifically on the questionnaire. The semistructured interview format allows the interviewee to probe further where appropriate. The value of the semistructured interview is in the ability of such a format to provide agency to the actor. Touraine (1988) argued, "The object of sociology is to explain the behavior of actors by the social relations in which they are placed. It is the relation, not the actor, that we must study" (p. 8). In order to explain the social relations of the actor, it is necessary that the interviewees themselves express these relations, and this is possible when the questions are not entirely predetermined and have the flexibility to allow the interviewee to determine the narrative flow of the interview process. Since the object of the study of *global connections* is to explain the strategic behavior of the subject population, I wanted the entrepreneurs to themselves explain to me what their strategic behaviors consisted of.

Furthermore, the semistructured interview is appropriate for gathering the kind of data required to fulfill my study. Leung points out, regarding her own research on Taiwanese entrepreneurs in Germany, that she used semistructured interviews to understand the multiple facets of entrepreneurs' lives, and to place their actions at a particular time and place. By participating in communal events and asking questions about the complex processes that underlie business decision making, the researcher states that she was better able to analyze "their seemingly 'pure economic' or 'pure ethnic' decisions" (Leung, 2001, p. 291). The deployment of semistructured interviews proved to be a very useful data-gathering instrument on Indian entrepreneurship in Silicon Valley since I was able to ascertain the links between the economic practices of immigrant entrepreneurs with their non-economic decisions. By training myself to use probes, I was able to get the interviewees to discuss issues that they deemed pertinent. Since the research question was intimately tied to understanding to what extent the entrepreneurs were global agents, the interview process allowed the agency to return to the actor.

RECRUITMENT

As stated earlier, I used my presence at the local associations, organizations, and meetings in Silicon Valley and Bangalore to recruit participants into the study. In Silicon Valley, I also relied on a local trade magazine, *Silicon India*, to provide me with a list of Indian-owned hi-tech companies. In Bangalore, I used NASSCOM listings as a guide.

Silicon India is a magazine (it has a separate web edition) with offices in Silicon Valley and Bangalore. It is well read in the technology community. It publishes a yearly list of top 100 companies run by Asian Indians in the United States, and as I found out in the course of my interviews, entrepreneurs go to great lengths to ensure that their ventures are included in the list. The companies are in five categories: telecommunications/networking, software products, software services, Internet/e-commerce, and semiconductors/EDA (electronic design manufacture). The publication is the best means of obtaining up-to-date information, especially given the turbulence in the industry with regard to the folding up of start-up operations or companies being bought out by competitors. The list includes Indian-owned companies across the nation, although, as expected, a large share of these companies is located in Silicon Valley. I focused on those companies that were listed under the software services/software products categories, with the expectation that these would be most likely to have connections with the Indian software industry.

NASSCOM is the National Association of Software and Services Companies in India. I used the directory to get a list of companies in Bangalore. In addition, wherever appropriate, I visited the Bangalore office of those companies that were based in Silicon Valley. The initial contact with the entrepreneurs was made through e-mail. However, the response rate was usually low. In the end, the best approach was through personal contacts established in the course of the fieldwork stay through participation in local events of the community. I also adopted the snowball method, which depends on individual informants to refer the researcher to other informants (Howard, 2002). Critics point out that the snowball technique can introduce bias into the overall sample. However, the bias would have been overcome by the fact that the initial personal contact was established through different associations. This is because the people who frequented TiE meetings were not the same as those who attended TANA events.

Treatment of Data

On my very first interview, I was taken aback when asked to sign a nondisclosure form by my entrepreneur interviewee. When I explained to him that the purpose of the interview was for academic research, he relaxed and explained that he had to be careful about competitors stealing his ideas, since the company was still at a very nascent stage. His apprehension made sense—the interview was conducted at the peak of the "dot-com" years, and such fears were not entirely misplaced. As a result of the initial interaction, I decided not to use the tape recorder I had carried for the interview. In that and subsequent interviews, I took handwritten notes.

For the data analysis, responses were coded for core categories, which created a descriptive overview of major trends in responses. The transcription was first coded in its raw form—the question followed by the answer. The transcribed data were then reviewed for common themes. This was followed by a more descriptive model of transcription, which consists of a summary description followed by illustrative quote. These summary descriptions were used for interpretation. Also, I took notes not just during the interview process, but also while immersing myself in observational settings in the field sites. Note-taking of mental observations and situations is a valuable ethnographic research tool (Emerson, Fretz, & Shaw, 1995).

A final concluding remark on the question of the validity of findings derived through ethnographic methods is in order. According to Brewer (1996), "ethnographic research is both a methodology and a method." (p. 204). It is not only a means of data gathering, but also a whole approach to the research process, that of interpretive sociology. The cornerstone of interpretive sociology is Weber's definition of sociology: "a science concerning itself with the interpretive understanding of social action and thereby with a causal explanation of its course and consequences" (Weber quoted in Giddens, 1971, p. 146). Furthermore, for Weber, interpretive sociology attempts to explain the actions of the study population in the words of those experiencing the phenomena. This has led to a bias against qualitative sociology that dismisses such research as "anecdotal" or "documentary journalism" (Dickson, 1996, p. 199). However, the literature has long documented ways of doing systematic qualitative research (Brewer, 1994; Hammersley, 1989, 1990, 1992; Stanley, 1990). Moreover, feminist scholars have also pointed out that interpretation and empathetic accounts are a means of overcoming the androcentrism of positivist methods (Reinharz, 1992). More generally, ethnographic methods have been seen as an alternative to the false claims to objectivity of other research methods.

ENDNOTES

1. Geertz's methodology entailed immersion into the field site rather than assuming a distanced objectivity. His interpretive method involved understanding through observation and recording, but also interpreting signs and symbols used by people to analyze cultural practices.
2. The scale of activities of national agents remains at the national level, but the operations of transnational and global agents regularly occur across national boundaries.
3. BJP stands for Bharatiya Janata Party, a Hindu nationalist political party in India that was in power from 1998 to 2004.
4. Personal interview with Anil G., entrepreneur in Silicon Valley, January 2000.
5. The discussion on global forces, global connections, and global imaginations is based on the review of the literature in Gille and O'Riain (2002).
6. Bangalore is located in the southern Indian state of Karnataka. The city has a population of 6.5 million, with an additional 2 million living in the outskirts.
7. SEI (Software Engineering Institute) CMM (Capability Maturity Model) is a model for the interpretation and application of Total Quality Management in the area of software development. Level 5 is the highest rank an organization can aspire to.
8. NASSCOM is India's National Association of Software and Service Companies, the country's chamber of commerce for the IT software and services industry.

Chapter 5

Middle-Class Path to Entrepreneurship

Education, Family, Networking

The transnational immigrant Indian entrepreneurs in Silicon Valley exemplify not only the power of knowledge-based industries in the development of global capitalism, but also the increasing incorporation of actors from the semiperiphery who have the capacity to, at once, leave an imprint in two different geographically separated zones. This chapter and the next, together, weave a portrait of the new Indian "immigrant"—chapter 5 cracks open the shell to reveal the kernel of factors that allowed previously risk-averse, salaried, middle-class immigrants to uncharacteristically opt for setting up wealth-generating business ventures. Chapter 6 examines the present transnational practices of this capitalist class.

The three pillars that support the entrepreneurial base of hi-tech Indians in Silicon Valley are education, family, and networking. Their educational background allows Indians to fit into the prevailing cultural ethos of the Valley, and equally, separates them from other immigrants who also engage in self-employment, albeit in low-skilled sectors. The family support highlights the invisible role of women behind the successful entrepreneurial men. Finally, what began as informal networking, setting the stage for the first entrepreneurs to find a steady footing, has matured and emerged as a powerful tool for successive generations of entrepreneurs.

EDUCATION: BUILDING A STRONG FOUNDATION

Education was the basis upon which the migration pattern of "brain drain" emerged, and several of the entrepreneurs interviewed immigrated to the United States as part of the "brain drain" labor flow. In fact, of the 26 entrepreneur interviewees in Silicon Valley, 9 arrived in the United States in the 1970s, 16 in the 1980s, and only 1 arrived in the early 1990s. Furthermore, not surprisingly, of the 40 entrepreneurs at both sites, 9 had doctoral degrees, 18 had a master's, and the remaining 9 had bachelor's degrees, all in various braches of either engineering- or computer-related subjects. The high education levels would translate—in the period of brain drain of the 1960s, 1970s, and 1980s—into well-paid white-collar jobs. None had taken the path to entrepreneurship immediately upon entry. Entrepreneurship entered into the collective imagination of Indian software professionals only in the 1990s.

The typical trajectory was to gain admittance as a student and then attain employment in a professional capacity. None of the interviewees indicated that the desire to migrate was to become an entrepreneur. The primary reason stated was to attend a U.S. university and/or to seek professional advancement.

The shift from nonentrepreneurial to entrepreneurial activities, a trend that began in the 1990s and has increased in the decade since, takes on sociological significance for the following three reasons. First, there are traditional cultural barriers to business activity, where self-employment is viewed with disdain and seen largely as an activity passed on within families of trading castes in India. Second, entrepreneurship was not viewed as an option because of the emphasis on education (particularly education in engineering and sciences) and the rewards that this education brought—professional stability and a secure income, the two elements of respectable standing in society.

Finally, it is only within the context of entrepreneurship that the power of the educational elite has emerged—previously, when technically elite immigrants were nonentrepreneurial, their engagement with the country of origin was minimal, limited to the ability of sending remittances. Now, through their entrepreneurial activity and their ability to conduct global business activity, the entrepreneurial class is not merely helping change how the American hi-tech sector works (as seen in the rise of outsourcing), but is also influencing how India engages with the global economy. The issue of the power of the new transnational entrepreneurial class will be explicated in greater detail in the next chapter. That this "new" India is emerging out of a spirit of entrepreneurship—especially from the ranks of the previously risk averse—confounds expectations of scholars who have viewed India as being economically stagnant in the face of traditional strictures of caste and religion; I include here scholars such as Weber (1946), Barrington Moore (1966), Louis Dumont (1980), and Samuel Huntington (1993). I shall return to these three points in the course of this chapter.

The immigrant history of Anil C. highlights not only the trajectory from student to professional to entrepreneur but also how the barrier to entrepreneurship was not merely a financial one, but also one of overcoming cultural stereotypes about what it means to be self-employed. Anil C. described himself as having grown up in a "very, very middle class household," in India where "studying hard, doing well in school and landing up in university to study engineering was drilled into me everyday."[1] None of the interviewees indicated that they were from either "upper-class" or "lower-class" backgrounds. The entrepreneurs in the sample were clearly middle class in origin—19 entrepreneurs indicated that their fathers were central government employees,[2] whereas 12 said they held professional positions as engineers, doctors, teachers, or lawyers (these entrepreneurs were more likely to indicate that they were from "upper-middle class" families), only 2 indicated that their fathers ran their own business. A large number of entrepreneurs (28 of the total 40) said that their mothers were housewives.

Understanding Anil C.'s family, educational, professional, and entrepreneurial history allows us insight into a typical path of an Indian immigrant entrepreneur. Having completed his bachelor's degree in Electrical Engineering at the University of Roorkee[3] in India in the late 1970s, he arrived as a master's student in Computer Science at a university in the Northwest of the United States. In 1982, he began employment at a wireless networking company in Massachusetts and was later transferred to the company's Silicon Valley office in 1987. He rose up the ranks during his tenure and became a manager in 1990. In 1994, he decided to partner with one of his colleagues, a non-Indian, and together, they founded a start-up in Santa Clara in Silicon Valley. The company continues to produce profits and went public in 2000.

Asked when he decided to shift economic orientation toward entrepreneurship, he replied,

> It was not because I didn't like the job I was doing. I didn't come to the U.S. to become an entrepreneur. The entrepreneurial bug didn't bite me until my soon-to-be business partner raised it as a possibility. He said—what do you have to lose? Let's give it a try for a year or two, if it doesn't work out, you can always go back to your job.[4]

The two impediments in the change from professional to entrepreneur were, in the entrepreneur's own words, "financial" and "cultural." The former is understandable; the latter deserves greater attention. The fear of financial loss was put to the rest with the possibility of being able to return to salaried employment if the venture failed. It is interesting that half of the nonentrepreneurial Indian executives I interviewed indicated that they had been involved in entrepreneurial ventures at some point in their careers, with varying degrees of success. Some never received funding for their business plan; others started, failed, and returned to full-time employment; while a third possibility was having seen a company grow and eventually being bought out by a competitor, most likely at an immense profit for the entrepreneur. There is a revolving door through which Indians in Silicon Valley are able to move in and out of entrepreneurship and full-time employment. The ability to use salaried employment as a "fall-back" further assists Indians in overcoming fears of financial harm.

To return to the issue of cultural barriers, what the entrepreneur meant when he referred to "culture" was this: "There was never any culture of business in my family." He asked if I understood what he meant by the difference between "business" and "non-business" families in India. Even though I thought I did (this is one of the problems of being an "insider" when conducting field research), I wanted him to explain it to me in his own words, and he responded:

> Business families in India don't care about education. Why bother when you know you will inherit the family business. That was not the case for me. I come from a simple family—we were taught to learn, gain knowledge… yes, knowledge and learning were very important. I guess that is what Brahmins do—they go for education. I am not saying I believe in this caste stuff, I am open-minded about these things, but I did grow up thinking there was a difference between studying and doing well professionally and going in for business. Business in my mind was for people who didn't care about learning, only making money and thinking about profits.[5]

Another Indian CEO stated:

> We [Indians] didn't come here to become multi-millionaires. The thought never crossed our minds. We came to get a higher education in the best universities in the world. We came to get better value for our education. I would never have thought that I would end up being a co-founder of 3 companies in 10 years.[6]

The same CEO further said:

> Most of us guys here are from middle-class backgrounds. Take my father—he was a government employee all his life, very hard working, but he would never have considered becoming a businessman. Business is for *banias*,[7] that's what he would say. Entering business was definitely looked down upon. My siblings and I were taught that we

> had to get a good education. When I got into IIT [Indian Institution of Technology], he was thrilled. He said, "Your life is set." And that is how I felt too. I came to the US as a graduate student and thought I would earn a better living here. I saw myself, and many of my friends, become managers, high-level executives. Then Silicon Valley started going crazy with start-ups and some of us thought, why not us? We are just as good as they [natives] are. We have the knowledge, the skills, some of us have earned well and have good savings for seed money, why not use it towards something else? I would say that Indians became...I don't know...more confident. And then once the early ones did well—the Vinod Khoslas, the Kanwal Rekhis,[8] then it was like a flood...If there was no Silicon Valley, there would be no Indian multi-millionaire entrepreneurs in the U.S.[9]

The entrepreneurs had to reconcile the ways in which they perceived the values that they upheld (those of education and learning) and those they were uncomfortable with (engaging in money and profit making). The cognitive disjuncture was eventually resolved by understanding that their business ventures were built on educational training—being an entrepreneur in the information technology sector was not only about profits but also being schooled and well versed in the field of technology. Anil C. expressed how his educational credentials assisted his entrepreneurial aspirations:

> [The reason I say that] education was important in my founding the company is because even though my own educational and professional background did not teach me the skills of doing business, it taught me to be confident in my technical abilities. I am able to get funding not only because of my business plan, but also because of my resume. The fact that I had done well professionally, I think, was important. It gives investors more confidence.

> I think it is okay to try your hand at business and fail—VCs [venture capitalists] might even look at failure as something good, a learning experience, but if you don't have the basic skills, you are nothing in this industry.[10]

The key to analyzing the entrepreneurial drive among middle-class Indians is to understand it as being simultaneously steeped in the acquisition of technical skills through higher education and the culture of entrepreneurship that Silicon Valley was based on. Hence, to return to a point raised in chapter 1, it was not just the entrepreneurial culture of the Valley that immigrant Indians adapted to: Implicit in the quest toward higher monetary returns was the understanding that self-employment could not be achieved without first attaining the requisite educational and professional skills.

Interestingly, half the entrepreneurs indicated to me that their businesses were not in their areas of educational expertise (although this was less the case among those entrepreneurs with doctoral degrees). Far more entrepreneurs indicated that their professional careers and business activity matched, and that they were using the skills obtained through work experience. However, even when there was a dissonance between educational training of the entrepreneurs and their chosen area of specialization in self-employment practices, they nevertheless stated that education was an important value to uphold. Gaurav D., a software entrepreneur, stated that

> I was never one of those first-ranked students in school. My parents were always worried about me. So I am not one of those typical Silicon Valley types—IIT engineering, Stanford MBA, no, nothing like that for me. I managed in school in India, came here for an MBA—at a small, no-name school in the mid-West. Then I joined a company in New Jersey, learned the ropes, and with my business partner, came to

> Silicon Valley and we started a company and we're doing really well…But y'know what—I tell my kids now, don't take school lightly, because I won't allow my kids to just sit around, do nothing, and just inherit money from me. A good education is absolutely necessary. My son tells me—"Bill Gates was a college dropout." And I give it right back—"And he [Gates] regrets that decision." [Laughs] Maybe it is all the highly educated Indians working in Microsoft that helped him [Gates] change his mind about the importance of education.[11]

A similar emphasis on education was reflected in the interviews conducted in Bangalore. Shyam was raised in a small town in central India, where his father was a government employee and his mother a housewife. After one unsuccessful attempt to pass the entrance examination for IIT, he tried a second time, and was able to enroll in the Madras campus. He noted this fact with some degree of pride. "It was important for me only to go to IIT, even if it meant that I lost a year." He moved to Bangalore in 1989, and began work at a multinational company. In 1995, he set up his own software business in Bangalore. Asked why he thought education was one of the key elements for a successful entrepreneur, he explained: "When you build a house, you need a solid foundation. Education is that foundation for life. I don't see myself as a businessman, I am an entrepreneur." Intrigued, I asked the difference between a "businessman" and an "entrepreneur," to which he responded:

> For me, being an entrepreneur first starts with having an idea. In the software industry, you think about how you can convert your technical idea into something that is worth marketing. Businessmen start with the idea of how to make money; it is the money-making that is exciting. The image you have of a businessman is somebody who will grease palms, do corrupt things to keep his business going.

> Entrepreneurs are first excited with understanding how things work. For that, you need to be educated. My mind is an engineering mind—I want to take things apart and understand the mechanisms behind it. Then I want to know how this can be converted into a business model. Of course, once you start a business, you want it to be profitable.[12]

Shyam also drew an interesting comparison between himself and two other prominent business tycoons in India, and the way he aligned himself to one, and distanced himself from the second, was important in his explanation for what it meant for him to become an entrepreneur.

> I admire Dhirubhai Ambani, but I cannot be him, even though we are both from small towns. He never even went to university, but made enough money to send his son to Harvard Business School. But the point is, that guy doesn't need to get an MBA. To me, that is just for show. My role model is Narayan Murthy; he's an IIT-ian and an engineer like me. He is somebody who I think is an entrepreneur, not just a businessman. He educated himself, and used that education to do something for himself, and for society—look how socially conscious he is. Even his chauffeur has stock in the company.[13]

Narayan Murthy is the founder of the highly successful information technology giant, Infosys, headquartered in Bangalore. Always ranked among the top three IT businesses in India, the founder is the role model for the aspiring entrepreneur because of a shared common background in engineering. Ambani—founder of Reliance Industries, one of India's most successfully run enterprises, with interests in several different sectors of the economy—was admired equally for his rags-to-riches story, but not for the model to be emulated. The clear difference is on one individual's ability to make money using only "business" skills and the IT entrepreneur's more desirable combination of technical education, a social consciousness, and an entrepreneurial drive.

Another entrepreneur in Bangalore[14] surprised me by saying that his parents, who lived in far-away North India, were not aware that he was the CEO of his own company. At the time of my interview, he was 6 months into his new start-up and was waiting for an opportune moment to "clarify" to his parents that he was not the employee, but the owner of the company he worked for. He acknowledged that his was an "extreme" reaction, and he did not think he was being dishonest, just "careful not to make them (the parents) worry." He emphasized that it was not because being self-employed was in any way shameful, only that the parents would not be able to understand why the risks associated with running one's own venture was preferable to the stability of a well-paid professional job. He added that there were enough media reports of "people like him" starting businesses in Bangalore for him to worry that his parents may not approve. They (the parents) may have been upset if he had made the decision 5 or 10 years, ago because his father might have thought that he was "wasting an education" and becoming "too money-minded." With success stories of techno-entrepreneurs filling the pages of the daily press, there was no stigma attached to engaging in profit making. Therefore, according to this entrepreneur's logic, as soon as the company got off the ground, ensuring the ability to project stability over risk, he would inform his parents about his decision to turn entrepreneurial.

From the vignettes presented in the previous text, it is clear that attaining an education, especially an education in engineering, is considered highly advantageous. Equally importantly, a foundation in engineering presents the possibility of a secure and stable livelihood. This is further evidenced by the fact that the entrepreneurs interviewed in Silicon Valley began their careers with little thought of eventually becoming self-employed, and oftentimes had long careers as successful professionals before entering the self-employment market. The software industry has opened avenues such that immigrants of Indian origin in the United States as well as Indian nationals in Bangalore now have the ability to use their human capital and convert it into greater financial rewards. The issue of increased financial success is important to note, because the idea of becoming extremely wealthy (as many of the Indian entrepreneurs now are) through skills in professions that previously only allowed for middle-class status is something very new in India.[15]

Furthermore, as pointed out in chapter 2, modernity in India is equated with scientific rationality. Pursuing engineering, among the most rational of disciplines, was vaulted to a position of esteem. When some pioneering Indian engineers, both in Silicon Valley and in Bangalore, were able to demonstrate their equal ease at being entrepreneurial, it emboldened others to follow in their footsteps and erased the taint associated with business practices. By putting the engineering degree to use in the pursuit of business, the software industry has allowed an entire generation of otherwise risk-averse, business-shunning, middle-class population to become the unlikely bearers of India's entry into the global software market and generate wealth and jobs for the country's growing numbers of technologically trained professionals.

Family as Support System

A second aspect of entrepreneurial success that emerged repeatedly in the interviews was the role of the family. To quote an entrepreneur in Silicon Valley, "you shouldn't overlook the role of the wives of all of us entrepreneurs from India, the sacrifices they are willing to make, only Indian families provide that stability."[16] When asked if there was something specific about the "Indian family" that distinguished it from families of other groups, the entrepreneur engaged me in a long discourse on how the Indian marriage system was particularly geared toward the uncertainties of business life. Here is an excerpt:

> When you are running a business, especially in the beginning, there are times when you are working up to 20 hours a day. In my first year, it was like that all the time, nonstop. My wife never complained. She told me not to worry about the kids because she was there for them. She was also working at the time, it wasn't easy on her. I can't say if Indian families are special, but just look around you, how many guys are divorced? I've been married 20 years, my [business] partner [also an Indian] has been married for about the same number of years. What I am saying is that you HAVE [emphasis mine] to write in your research about the strength that we entrepreneurs get from the stable marriages we all have. You don't just leave your husband or wife because things are tough. My wife understands why I am away from the house. And she has a life of her own. I think Indians approach marriage differently, I just think it is more stable because we understand the pressures; we make adjustments, compromises more readily. We both know we are in this together, and it [the business] would not work if it wasn't for her supporting me 100%.[17]

While no other interviewee was as emphatic or engaged me on the subject for as long, the importance of a stable family life was highlighted in several different interviews. Thirty-one of the interviewees were married with children, eight had never been married (all under the age of 30) and only one indicated that he was divorced. No married individual indicated being childless. The two women entrepreneurs likewise credited their husbands in helping out in the care of the children while they worked long hours in running their businesses.

An Indian venture capitalist in Silicon Valley intimately associated with TiE, the networking association for entrepreneurs of Indian origin, explained to me two ways in which Indian family life is suited to American business life.

> It is not something we [venture capitalists] look for, at least not explicitly. All I care about is somebody's business plan, their idea, the management team and their experience. Nobody cares about your personal life. But if you ask me, that personal life is very important for success. Business life puts a lot of pressure, and the successful entrepreneur is one who can weather those stresses. I think, with Indians, the family is like the backbone supporting you, you know if you have a stable life outside your business, it is more likely you will function better inside your business. I've seen two kinds of guys—one is where the wife gives up her job and supports the husband in the business but the younger ones are happy that their wives work. Many of these young women are professionals in their own right. Their [the wives'] incomes become the primary source of income while the husbands are busy in the initial stages putting together their business plans and not generating any money for the household.[18]

Unlike the parents of the entrepreneurs (many of the mothers were stay-at-home mothers) among the entrepreneurs I interviewed in Silicon Valley and Bangalore, this was hardly the case, especially for married entrepreneurs who were under the age of 35. Many of the women are software professionals themselves, with well-paid jobs. The industry continues to be male dominated, as seen in the fact that there are only a handful of Indian women entrepreneurs in the IT industry, and female students make up a very low percentage of total enrolments in the IITs. However, there is, increasingly, a rise of women in professional ranks, with women now making up about 25% of the workforce in some IT companies in India (Chakravorty, 2006).

Women are also an important element in maintaining the cultural life of Indians in the Valley. In one such organization, TANA, I was able to witness firsthand how traditions are maintained while simultaneously adapting to the impulses of living a "modern" life in the United States. Women members of the organization, some the wives of IT professionals in the Valley, arranged for traditional dance and music lessons for the children of professionals. What struck me was that these often took place after school or over the weekend, when the mothers would drop off the children for a number of hours while they attended to other tasks. It was an effective babysitting arrangement while at the same time allowing the children to keep in touch with their Indian roots. At another event organized by the association, the subject for discussion was "Hinduism and Business Life," where the speaker addressed how religion can commingle with business practices, and the message conveyed was how the idea of wealth generation is noble if it coexists with the tradition of charitable giving.

Thus, although women may not be at the forefront of the changes within the IT industry (especially in the most visible aspect as entrepreneurs), nevertheless, those within the industry express the importance of the role played by women, whether as equal breadwinners or as a crucial support system that provides a critical source of stability in times of the inevitable stresses associated with turning entrepreneurial.

From Informal to Formal Networking, From Silicon Valley to India

The final aspect of entrepreneurship is the importance of networking in business circles within the Indian IT community. The lines I trace, with the help of interviews and observations, show the change, to use Marx's famous words, from a "class in itself" to one that became a "class for itself." The first adopters—those who went from professional work to a highly remunerative career as entrepreneurs and became the first immigrant Indian IT millionaires—relied on informal sources of networking to set up their companies. Realizing the potential of those like them—professionally qualified Indian immigrants in technology fields—these pioneering Indians endeavored to have others join them in their ranks by setting up formal networking associations. Meanwhile, Indians in Bangalore were likewise beginning to awake to the potential that the industry offered. Together, these two groups of Indians have formed a formidable team of entrepreneurs and professionals who are heralding a new group of very wealthy Indians. Indians in Silicon Valley are now able to give back to their country of origin, as we shall soon read, in both material and nonmaterial ways.

A longtime resident of Silicon Valley narrated to me how he set up his company in the late 1980s. An IIT-Delhi graduate, he arrived in the United States in 1971, obtained a doctorate in Mechanical Engineering in 4 years at a prestigious Northeast university, and worked in a professional capacity for the same company in Silicon Valley for 11 years. At the time he left the company, he said his salary was "well over $100,000 a year." Having enjoyed professional success, he wanted to try his hand at forming his own business. Where did the idea emanate from?

> I looked at my peers—this was the mid-1980s—and it was common for people in my team to leave the company and set up on their own. This was the story in Silicon Valley in those days. It still is. You work, you have your own ideas, you leave and become an entrepreneur. I saw people leave and become extremely successful, become

> very wealthy. I didn't see any Indians doing this at the time. I thought about it for a long time. I mentioned it to a couple of other Indian friends of mine, their reaction was not encouraging. Finally, one day at a dinner gathering with some friends, I mentioned it again. One guy, a very successful Indian cardiologist friend asked—"how much money would you need?" I just said casually—"Why, will you fund me?" To my surprise, he said, seriously, that he would. So I pooled in some of my resources, got money from my friend, and I came up with a business plan. There was hardly any paperwork we signed—he just wrote me a check. I didn't go through the formal VC [venture capital] channel because I wanted to prove something before I got "official" funding. The business took off and I was able to return the initial investment to my friend and more. In the second round of funding, I approached a VC firm and by then, I knew what I was getting into. Had the lawyers, got the paperwork straight. The Indians starting now have a lot to learn from the early birds like us.[19]

Professionally qualified Indians—the doctors and engineers—who had immigrated in the period of "brain drain" became "angel investors" for the early Indian immigrant entrepreneurs. Angel investors are so called because they are informal investors from within the circle of family and friends and are thus differentiated from the formal venture capital firms. That some Indian immigrants were in a position to become angel investors emerged from the wealth they were generating in their professional work, a fact that has come to light by the year 2000, as the latest census shows that Indians are now the highest income group in the United States (U.S. Census, 2000). Some might argue that relying on informal sources of funding is akin to the transactions conducted by other immigrant entrepreneur groups, such as the use of rotating credit associations among Korean immigrants that have been noted by scholars (Light & Bonacich, 1988).

However, relying on wealthy angel investors is not atypical in the Valley and can be discounted as a practice that is particularly ethnic in origin.

The idea to formalize networking and provide an avenue for future entrepreneurs of Indian origin germinated among those who became successful in the 1980s. By the mid-1990s, Indians in Silicon Valley had their own networking associations that they could rely on, and the next chapter will highlight the power of one such group, TiE.

Among the entrepreneurs I interviewed, only three in Silicon Valley said that they started their companies using TiE networks, but all acknowledged that the formation of Indian networking associations has other benefits. These include the ability to meet like-minded people and discuss business issues, but also, interestingly, it was pointed out that TiE allows visibility for Indians among other non-Indian groups in the Valley, including venture capital firms who now see Indians as being able to run successful companies.

The connections that Indians in the Valley are able to create with their country of origin deserve some attention. There are different ways these connections are being enacted—in material ways when immigrant Indians set up subsidiaries of their companies in India, generating employment for the local population, or through the donations made to charity organizations and to their Indian alma mater. Regarding the latter, the sums under consideration are not insubstantial—to give just one example, Gururaj Deshpande, Indian immigrant founder of the highly successful Sycamore Networks, has pledged $5 million every year for 20 years to his alma mater, IIT Madras (Naroola, 2001). If these were rare and episodic cases, it might not merit attention, but all seven IIT campuses in India have received similar sources of funding from their alumni in the United States.

Anil C. spent some time telling me about his decision to locate one of his subsidiaries in Calcutta, a metropolitan city in East India, a city not known to be on the forefront of the Indian IT industry map. However, when the time came for the entrepreneur to decide where to locate his second off-shore facility (the first one is in the growing IT hub of NOIDA, a suburb of New Delhi), he chose Calcutta. He indicated that his decision was purely based on a need to connect with his home city, and it was not as though "there aren't trained people in Calcutta." In doing so, he hopes to be able to "help India and Calcutta in some small way." The company now hires 150 employees at the Calcutta office.

The nonmaterial ways of assistance offered by immigrants lie in their hopes of bringing in Silicon Valley style networking meetings and gatherings to help mentor young entrepreneurs in India. It is unclear to what extent these are successful, and it is beyond the scope of the data. However, it is safe to say that the membership of IT associations in India is on the rise. Finally, there is a transmission of management styles between these two locations—Silicon Valley and Bangalore—not least helped by the fact that Indian IT executives in the Valley have assisted their American employers in setting up offshore facilities in India. One such interviewee, an employee of a large multinational company in the United States, explained to me that she was chosen to set up the Indian facility, and that she sees no difference between the American and Indian facilities. Even in the way the buildings are constructed in Bangalore's Electronics City, home to the sprawling Infosys campus among others, one observes the similarity with the architectural styles of companies in Silicon Valley.

Thus, the connections that are drawn between these two locations are numerous. In addition to demonstrating the rise of middle-class Indians to positions of wealth through the practices of entrepreneurship, the chapter has also highlighted how this class of diaspora Indians is enabling changes within their country of origin. The change in India, from an autarkic economy based on a socialist model, to an economy at the forefront of the global IT industry, is contrary to decades of scholarly writings on the inevitability of India's economic sluggishness. What makes the change more startling is that it is being led by traditionally risk-averse middle-class Indians, who have transformed India's position in the global software market within a short span of time. Weber, as pointed out in chapter 2, argued that the Indian caste system was a barrier to the country's entry into Western-style capitalism. Barrington Moore used India as an example of a country where political enthusiasm for democracy was not accompanied by a vibrant bourgeoisie. In the present stage of global capitalism, middle-class Indians, at least in the information technology industry, are an important force in creating a vibrant environment of trade, enabling India's entry into the global software industry. Indians who were formerly middle class but have now risen to positions of wealth and power would like to see India's image change along with their own changing fortunes.

The following chapter looks at the power of this newly emerging techno-entrepreneurial elite and its implications for how one should analyze the transnational practices of highly skilled Indians and differentiate these groups from low-skilled immigrants, the latter forming the bulk of scholarly literature on immigrant entrepreneurship, and immigration in general.

Endnotes

1. Personal interview with Anil C., entrepreneur in Silicon Valley, June 2001.
2. The central government bureaucracy in India ranks employees into four categories—Class I through Class IV, Class I being the highest in terms of both pay scale and status. Class I employees are selected on the basis of written and oral examinations and undergo training for their positions. Although some Class I employees rise up the ranks to become senior-level diplomats—for instance, ambassadors or high commissioners to foreign nations—none of the interviewees indicated that their fathers had reached these positions. Most entrepreneurs who answered that their fathers were government employees answered Class I or II, only 1 said Class III (typically, a clerical position), and none said Class IV (usually, the category that includes peons and other low-level jobs). It can be safely argued that government employees in India are good examples of the core of middle-class India.
3. University of Roorkee is now one of the Indian Institutes of Technology (IIT); see table 3.
4. Personal interview with Anil C., entrepreneur in Silicon Valley, June 2001.
5. See endnote 4.
6. Personal interview with Akash N., entrepreneur in Silicon Valley, January 2000.
7. *Banias* are the traditional caste of traders in India.
8. Both Vinod Khosla and Kanwal Rekhi are considered the stalwarts of the Indian community in the United States, and are widely known in India as well. They are among the earliest Indian entrepreneurs in the Valley, and were already running successful companies by the early 1990s. Vinod Khosla cofounded Sun Microsystems and is currently a partner at the venture capital firm Kleiner Perkins Caufield & Byers in Silicon Valley. Kanwal Rekhi started a highly successful company called Excelan, which was eventually sold to Novell Networks. Rekhi and Khosla started TiE, the Indus

Entrepreneurs, the networking organization that has helped fund several Indian-owned ventures. Both are graduates of IIT in India.

9. Personal interview with Akash N., entrepreneur in Silicon Valley, January 2000.
10. Personal interview with Anil C., entrepreneur in Silicon Valley, June 2001.
11. Personal interview with Gaurav D., entrepreneur in Silicon Valley, January 2000.
12. Personal interview with Shyam S., entrepreneur in Bangalore, April 2000.
13. See endnote 12.
14. Personal interview with Madhur, entrepreneur in Bangalore, March 2000.
15. The new wealth is not restricted to entrepreneurs alone. Even software professionals at top companies in India are experiencing monetary benefits that were unheard of before the arrival of the software industry. To provide another, more personal, example from the time spent during fieldwork: a high school classmate of mine is now married to an engineer who works for a top Indian software company. At the time of my meeting the couple in Bangalore, they were perusing through catalogs (itself a new phenomenon) to buy a new car and a new bungalow-style house in an upscale, upcoming suburb in Bangalore. While this piece of information may not be surprising to "outsiders," particularly in the United States, where young professionals are highly likely to start their careers by investing in these two amenities—house and car—this was still surprising, if not shocking, to my eyes as it was, until now, unheard of for individuals in their late 20s to have the ability to pay for a new car and home, which would have previously taken a lifetime of savings from middle-class pockets.
16. Personal interview with Subash T., entrepreneur in Silicon Valley, November 1999.
17. See endnote 16.
18. Personal interview with Sirish G., Silicon Valley venture capitalist, May 2001.
19. Personal interview with Pravin D., Silicon Valley entrepreneur, June 2001.

CHAPTER 6

TRANSNATIONAL CAPITALIST CLASS OF INDIAN IMMIGRANT ENTREPRENEURS

Transnational immigrant entrepreneurship has been viewed, and sometimes valorized, as a form of "globalization from below" (Henry, McEwan, & Pollard, 2002; Kyle, 1999; Mahler, 2002; Portes, 2000). In the following pages, I pose the question: Can the activities of transnational Indian immigrants in the software industry be likewise treated as a means of opposition to the dominant structures of global capitalism?

The argument presented by the proponents of the "globalization from below" thesis is as follows. International migration in the last half century has been fuelled by the twin forces of the global industrial restructuring of capital and the demands for labor in industrialized societies.[1] There are two ways in which widespread industrial restructuring has impacted migration. Firstly, loss of manufacturing jobs has created a polarity wherein high-skilled, highly paid professionals fill the primary labor market, while their lifestyle and consumption needs are met by a low-wage labor force. Immigrants are increasingly occupying these low-level positions (Cornelius, 2003; Hum, 2001), as an educated domestic workforce is disinclined to fill the low-paying service jobs (Kandel & Parrado, 2004).

Secondly, the global mobility of capital has also had consequences for the supply of potential immigrants. Capital is always in search of new countries in the underdeveloped world and it is usually middle- and lower-middle class households that are the most exposed to Western modes of production and consumption aspirations. These groups are "pre-socialized into the emigration process" (Grasmuck & Pessar, 1991). Thus, contrary to common public perception, it is not the poorest nations that are the primary sending countries or the poorest segments of society that are the most likely to migrate; rather, it is the working-class and middle-class sectors of intermediate-wage nations that are the most likely to move (Portes & Rumbaut, 1996; Sassen, 1998; Staring, 2000), converging in the "global" cities of the world, such as New York and Los Angeles (Sassen, 2001; Savage, Warde, & Ward, 2003).

Once in their destination country, finding their options to be limited, immigrants turn to various alternatives to secure an economic footing—entering the informal economy[2] (Kloosterman, van der Leun, & Rath, 1999; Valenzuela, 2003), increasingly migrating to rural areas of the United States to seek employment (Kandel & Parrado, 2004), or engaging in self-employment practices. Immigrant entrepreneurs find themselves faring better in the labor market, earning more than their waged coethnics, even controlling for human-capital variables (Logan, Alba, & McNulty, 1994). Thus, for a number of ethnic groups, entrepreneurship has been found to be an oft-repeated path of immigrant incorporation into advanced, immigrant-receiving countries (Light & Bonacich, 1988; Waldinger et al., 1990).

More recent literature on immigrant entrepreneurship has documented new patterns of immigrant adaptation, in the form of transnational business practices that involve regular and sustained economic activities that cross national boundaries (Itzigsohn, Dore, Hernandez, & Vazquez, 1999; Landolt, Autler, & Baires, 1999). These immigrant entrepreneurs are "neither here nor there, but at both places simultaneously" (Portes, 2000, p. 254). Further research has noted variation in the extent to which groups engage in crossborder economic linkages depending on individuals' human capital endowment, size of social networks, and context of exit and reception (Guarnizo, Portes, & Haller, 2002).

It is not just transnational economic activity that has been subject to scrutiny. Scholars have shown the rise of sociocultural transnational activity. Itzigsohn (2002) examined how immigrant organizations promote cultural and social ties with the country of origin; Golbert (2001) showed how Ukranian Jewish youth identity was shaped by transnational identities of the Jewish diaspora; and Guarnizo, Portes, and Haller (2003) argued against an overly optimistic view of transnational political activity, suggesting that when groups do engage in political fields across national borders, it often results in reinforcing existing power imbalances. The results have shown variation among ethnic groups by class, citizenship status, and gender, on the extent to which immigrants engage in transnational activities, but all wholeheartedly attest that transnationalism, along several different axes, is on the rise.

To take the argument back to where we started it, at its base, then, immigrant entrepreneurship is driven by the logic of capitalism itself, as both the process of migration and the choice to become an immigrant entrepreneur emerges within the context of global mobility of capital and the global need for labor. What is more, "because the phenomenon is fueled by the dynamics of globalization itself, it has greater growth and potential and offers a broader field for autonomous popular initiatives than alternative ways to deal with the depredations of world-roaming capital" (Portes, 2000, p. 255).

Thus, immigrants who engage in transnational entrepreneurship are using the same logic that determined both their decision to move and their fate upon arrival. In the same way that global corporations use the mobility of capital to restructure and downsize their operations, immigrant entrepreneurs have also learned the mechanisms behind portable capital and transformed them into their own versions of transnational business practices. In the same way that global industries depend on new communication technologies to succeed in their globe-trotting initiatives, immigrants too comprehend that the new tools at their disposal can indeed shatter their hitherto vast distance from their homeland to offer them a means of engaging with their countries of origin. By taking advantage of these new possibilities of entrepreneurial engagement, the migrants resist poor working conditions in host societies and create personal mobility. Activating their extensive crossborder social networks, transnational entrepreneurs give to themselves an alternative, more liberating way to overcome their economic situation. As immigrants are not part of the dominant, decision-making groups in society, since forces beyond their control (such as industrial restructuring) drive their actions, this form of transnationalism is referred to as "grassroots transnationalism," or, a globalization that is occurring from "below" (Portes, 1999).

A number of studies empirically demonstrate how this globalization from "below" is being enacted. Landolt et al. (1999) show that while traditionally, Salvadorans in the United States interacted with their home country by sending remittances, a practice that has increased over the years, the rise of transnational Salvadoran entrepreneurs has changed the relationship between sending and receiving countries. Salvadorans who regularly travel to and from their two "homes" are incorporating their countrymen into their business practices in the United States, so much so that Salvadoran banks and enterprises have started to invest in these immigrants businesses. When Henry et al. (2002) write about the entrepreneurial practices of the diaspora population of Birmingham, they argue that the Chinese, Indian, and Greek entrepreneurs are using their dense overseas networks to create innovative and unique business products that are "ethnic" in origin, but with an important difference. Even though these immigrants are selling products that have a distinctive ethnic stamp that allows them a competitive advantage over their native counterparts, what is *not* ethnic is their clientele.

Similarly, Kyle (1999) shows that traditionally, the indigenous Otavalo people of Ecuador have specialized in the production of clothing, but have been reliant on middlemen to peddle their wares around the world, taking in little of the profit. Under conditions of globalization, however, the Otavalans have realized the commercial value of their traditional crafts and have become transnational entrepreneurs themselves, moving back and forth between Ecuador and countries in the West, subverting the need for traders who were historically the intermediary node between these third-world indigenous producers and their Western consumers.

Nor is transnational activity limited to immigrants. Evans (2000) sees the possibility of a counterhegemonic, grassroots, working-class labor movement emerging as a result of cross-national alliances between unlikely partners, such as third-world employees in different countries of giant multinational corporations.

In all these instances where transnational business activities of immigrants have been established in the literature, and especially those that contest that these practices should be dubbed as transnationalism from "below," the image of immigrant entrepreneurs is one where the migrants subvert the dominant forces of capitalism, even when engaging in the same logic that underlies late capitalist development. The overarching theme is one of resistance to, and defiance of, the existing global world.

How does the idea of globalization from "below" translate when applied to Indian immigrant entrepreneurs in the hi-tech sector? Following a long period of brain drain of qualified personnel, which was a loss for India, the immigrants are now showing increasing propensity toward linking up with their country of origin. In this way, it is a new form of immigrant incorporation and economic adaptation. This is in keeping with the literature discussed in the previous text. However, the experience of Indian entrepreneurs is not a form of resistance to the dominant forces of capitalism, but is a complete embrace of it.

Although Indian software entrepreneurs in Silicon Valley are immigrant entrepreneurs because of their status as an ethnic minority group in the United States, their activities, and equally importantly, their discourse, is best captured by the term "transnational capitalist class" (Sklair, 2001).

Indeed, Indians in Silicon Valley are not very different from those in Bangalore, and the entrepreneurs, both within the national context of India and as immigrants in the United States, form part of a global class that aligns itself to the capitalist system we live in. In this way, far from being "a form of individual and group resistance to dominant structures" (Portes, 1999, p. 227), the transnational activities of Indian immigrant entrepreneurs in the software industry is instead an inextricable part of the dominant structure, propelling it, aligning with it, reveling in it.

Indian Software Entrepreneurs: A Transnational Capitalist Class

Sklair (1998) defines the term transnational capitalist class in a double sense: "[I]ts members have global, rather than, or in addition to, local perspectives and secondly, it typically contains people from many countries who operate inter-nationally as a normal part of their working lives" (p. 3). Furthermore, this class aligns itself with the capitalist global system and the transnational corporation (TNC),[3] and because the orientation of its members is outward looking: It opposes not only those who reject capitalism as a way of life but also those capitalists who reject globalization. The transnational capitalist class, according to Sklair, forms a global ruling elite.

Writing specifically about the transnational capitalist class in the Third World, the author outlines four features that distinguish the class in developing societies. (1) Transnational capitalist classes do not identify with any foreign country in particular or even necessarily with the First World, or the white world, or the Western world.[4] They identify with the global capitalist system; (2) they reconceptualize interests in terms of the global system; (3) they take on the political project of aligning their, and their conationals', interests in terms of the global capitalist system and; (4) the transnational capitalist class in the Third World may, under certain circumstances, begin to dictate its own terms to the TNCs and foreign interests. Let us turn our attention to how far the Indian entrepreneurs in Silicon Valley and Bangalore fit the characteristics outlined by Sklair in his definition of the transnational capitalist class.

The hi-tech entrepreneurs operate global companies, and conduct transactions internationally on an everyday basis. In India, software industry sales continue to remain export driven, with North America as the leading market, ensuring that although the companies operate in a specific local context, they are an embedded segment of the global software industry (NASSCOM, 2004b). All of the interviewees in Bangalore, including the entrepreneurs, executives, and employees, had experience working overseas. Of the 28 interviewees, only 5 were return migrants because these individuals had immigrated overseas and made the decision to return to their country of origin. However, the remainder had worked on short-term overseas projects, and thereby lived in foreign locales, from anywhere between 2 months to 2 years.

Similarly, in Silicon Valley, interviewees' professional lives existed within an international context. This was particularly true of the entrepreneurs and high-level executives, although less so for software professionals. Of the 26 entrepreneurs in the study sample, all had global operations, with 21 stating that their facility in India was a critical component of their business venture, and that they traveled to India on a frequent basis. The transnational movement of Indian-born entrepreneurs in Silicon Valley to their country of origin has been documented elsewhere in the literature (Dossani, 2002; Saxenian, 2002). Thus, Indian entrepreneurs in the software industry, whether in the United States or India, are deeply intertwined in economic relationships outside of their national or local bases, forming a crucial component of the global software industry.

Allegiance to the Global Capitalist System

The allegiance of the entrepreneurs, as Sklair predicts, is not to any specific national or local context. The identification is with the global capitalist system. When asked why he had decided to set up his company in Silicon Valley, Rajeev M., an Indian entrepreneur, replied,

> I am here because this is where I can run the best tech company in the world. I used to be an employee, now I am in the business of running a business, so you go where the environment can give you something. This is the environment for entrepreneurship, for venture capital funding. If I were into making toys, I would go to China. If I wanted to grow coffee, I would go to Brazil. Why am I in Silicon Valley? I'm a tech guy, that's why I am here. It's the best place to turn your technical idea into a world-class corporation. Because this is where the money is, the resources, the guys with who you can talk the lingo about hi-tech start-ups. I like living in California, but that is not why I am here.[5]

Probed further about the company's operations in India, the interviewee cut me short to correct me: "We are not just in India, we are all over Europe, Asia, Asia Pacific. It is important for people to realize that we do business all over the world, not just in India. We are a world-class, transnational organization." Later, he added: "Look, I go to India all the time. What we do there is important. We are going to set up an R&D facility there. I just want to make sure you understand that we do business all over the world."

The immigrant entrepreneur is motivated by the desire to run a "tech company," and the choice of location is determined by where the needs to fulfill that desire would most likely be met. In the same way that the logic of capitalism dictates the need for capital to be mobile, for the entrepreneur, the location of the business is secondary to the location of the capital required to run the business. Moreover, it was important for him to clarify for me that this was a truly global business, with international offices. As a capitalist in the technology sector running a "world-class, transnational organization," his presence in Silicon Valley, the world's leading center of hi-tech venture capital funding, is *de rigueur*. If he were a toy manufacturer, he would find himself in China, and Brazil, if he decided to be a coffee grower. The identification, thus, is not, as Sklair puts it, with any particular foreign country, or any location in the Western world. It is singularly in the interests of the location that best serves the interest of being a global capitalist in the hi-tech industry.

I do not mean to suggest that these entrepreneurs have no sentiments of attachment to their countries of origin and adoption. For instance, survey results have shown that when Indian professionals and entrepreneurs in Silicon Valley express an interest in migrating back to India, or becoming transnational entrepreneurs, "contribution to home country" and "culture and lifestyle" (see table 15 in chapter 3) rank high on the list of reasons given for the return decision. However, the nationalist sentiments and longing for "home" are also tempered by an understanding that there are now opportunities in India that were unthinkable even a decade ago. The concomitant rise of entrepreneurship in two locales, geographically separated, but united under the banner of the increasing globalization of the software industry, can also be seen through the lens of India's liberalization, which ended the nearly 50 years of autkarkic policies of self-sustaining, indigenous development.

India began to liberalize its economy in 1991, reversing 4 decades of post-Independence socialist-inspired policies of a large public sector, and extensive controls on the private sector.

The significant reduction in trade and tariff barriers allowed the growth of the software export industry (Heeks, 1996). In my interviews with entrepreneurs in both Silicon Valley and Bangalore, the identification with the capitalist system often took the form of decrying India's many social ills as a result of the country's state-owned enterprises, its socialist past, and hailing private entrepreneurship as an answer to the country's lack of development. More importantly, the entrepreneurs saw in their own activities the vision that the country had lacked for altogether too long. As one return migrant, Subroto B. put it to me in Bangalore:

> I spent many years in the United States before coming back to Bangalore to co-found a company here. India is different from what it was before. The reason I came back is because now there are opportunities here, it is no different than being in the States. I mean, there is no difference in what you can do in running a really good company in Bangalore. In Bangalore at least…I don't know about other places. Plus, I am closer to my extended family…but the other reason I am excited to be back here is because I feel I can make a contribution. I can actually do something to make a difference, change the system. All these government-run institutions are ruining the country. If people had to run something for a profit, they would not take things so easy. We should take the best from America…efficiency, work ethic, customer friendliness, business climate.[6]

The respondent speaks in glowing terms about the spirit of American capitalism and the core virtues of an environment that supports entrepreneurial energies, an environment that is now increasingly being incorporated into Bangalore's own hi-tech industry, making the decision to return to the country of origin a much easier one to make. Bangalore, per the interviewee, is now deemed no different from the United States, and the similarity is seen in terms of the possibilities of founding and running a hi-tech company in either location. What India suffered from before was a lack of private enterprise. Thus, inasmuch as the reason for return is because of the desire for closeness to the extended family, there is no doubt that what gives weight to the decision, and sets to rest the ambivalence of such a move, is the ability of easing into a capitalist system that one had familiarized oneself to, albeit in the context of another country.

I turn now to Sklair's second and third points in characterizing the transnational capitalist class in Third World societies: whether, and how, this class of entrepreneurial Indians is also using its economic power to realign political interests. Even though Sklair uses the term in the context of the Third World, as we will see in the text that follows, the applicability can be extended to Indian transnationals who have a vested interest in realigning ties between the sending and receiving countries.

Globalization as a Political Project of Transnational Entrepreneurs

The political power of Indian transnational entrepreneurs is discussed through the workings of TiE, an influential not-for-profit networking organization for South Asians in the hi-tech industry. In March 2000 U.S. president Bill Clinton issued a joint statement with the Prime Minister of India on the occasion of the president's official state visit to India. The statement read as follows:

> We are two of the world's largest democracies. We are nations forged from many traditions and faiths, proving year after year that diversity is our strength. There have been times in the past when our relationship drifted without a steady course. Globalization is erasing boundaries

> and building networks between nations and peoples, economies and cultures. The world is increasingly coming together around the democratic ideals India and the United States have long championed and lived by. Together, we represent a fifth of the world's people, more than a quarter of the world's economy. *We have built creative, entrepreneurial societies. We are leaders in the information age*. (emphasis added) (Joint India-U.S. Statement, 2000)

Relations between India and the United States were marred by decades of Cold War politics that pitted India's close ties with the former Soviet Union against the joint front posed by Pakistan's alliance with the United States. In the decade after the collapse of the Soviet Union, political ties between the two countries improved considerably, highlighted by President Clinton's state visit to India, the first visit by a U.S. president since 1978. The previous statement not only views India's interests to be coterminous with those of the United States because the era of globalization has forced open national borders, but also sees both countries sharing a common feature—each country has individually created entrepreneurial societies, becoming world leaders in the information age. What is not directly mentioned in the statement is this: The "relationship that drifted without a steady course" was set on its right path by the growing power of Indian transnational entrepreneurs in the software industry.

President Clinton was accompanied on his visit by a 200-person delegation from TiE[7], the South Asian networking organization for entrepreneurs, which was first set up in Silicon Valley in 1992 and now operates 40 chapters, including 24 in the United States and 13 in India, and has a worldwide membership of more than 10,000. Interestingly, TiE's mission statement emphatically denies that this not-for-profit organization has any political objectives, a statement made clear on its Web site. "Does TiE have any religious or political objectives? **No!** TiE does not engage in political or religious activities or any matters that may create divisiveness among people" (bold text and exclamation mark in original) (TiE Web site). This raises the question: What are the stated goals of an organization that is intimately involved in political relations between India and the United States, but in its presentation of self is vehemently nonpolitical?

One of the founding members of TiE expressed TiE's objectives thus:

> TiE was created by a small group of Silicon Valley entrepreneurs and professionals. We are strictly against divisions of caste, region, religion or language. That is why we chose the name Indus[8] Entrepreneurs, not Indian entrepreneurs, to be inclusive of all the peoples of South Asia, regardless of where in South Asia they come from, or what religious group they belong to. What binds us is our interest in wealth creation through entrepreneurship...Education is the leveler—we are all highly educated, so we don't have old divisions of caste and creed amongst us...We are a modern organization with a scientific outlook. The idea is to take the best of the two cultures we are familiar with—Silicon Valley's entrerpreneurial culture and India's traditional culture. See, India has a *guru-shishya parampara* [tradition of teacher-disciple relationship], which can be used for the idea of networking. Established entrepreneurs can teach and mentor budding entrepreneurs—that is the traditional Indian relationship between a teacher and his student. Then you blend that idea with Silicon Valley's economic culture of entrepreneurship. That was the idea behind TiE.[9]

TiE purports to be a blend of modernity and tradition. The modernity is partly avowed in its Nehruvian variant—the building blocks for a "modern" India rested on scientific and secular rationality. With a scientific outlook, it was believed, the divisions of caste, religion, and language would ultimately witness their own demise. However, for Nehru, these modern ideas for a progressive India could be combined with the spiritual tradition of an earlier time. "The scientific spiritual approach, and laying stress on the scientific part of that approach, has been the characteristic of the highest Indian thought, not your thought or mine but I repeat the highest Indian thought in the past ages… the two are essential." (quoted in Singh 1988, p. 180).

Through higher education in science and technology, Indians have imbibed these Nehruvian ideals, and thus education, as per Dr. K.'s statement, has been the great leveler, uniting all against the nonmodern, unscientific approach. Tacked on to these ideas of modernity are the lessons to be learned from ancient Indian spiritual tradition; for TiE, this comes in the form of the *guru-shishya parampara*, the traditional ties between the mentor and the disciple. In one critically important way, however, TiE's stated values are very different from Nehruvian thought. Nehru was a socialist, albeit a scientific socialist, whereas members of TiE adhere to the ethic of Silicon Valley, in all its global capitalist spirit:

> TiE endeavors to cultivate and nurture the ecosystems of entrepreneurship and free-market economics everywhere, as it sees this to be the single most powerful instrument of prosperity. The values and beliefs of TiE reflect the culture and value system of Silicon Valley: entrepreneurship and wealth creation are invaluable human endeavors, and successful entrepreneurs find fulfillment in helping budding entrepreneurs. (TiE Web site)

The intermingling of Indian cultural values with mainstream American business culture emerged in another unexpected manner. A non-Indian Silicon Valley venture capitalist remarked to me that

> Indians have strong family values. Their families are close-knit. This is important from a business standpoint. When you decide to set up a tech start-up, your time is consumed by the business. If you have a supportive family, it helps in two ways—one, it makes you more motivated because you want to provide for your family and secondly, you are less likely to have family problems which can adversely affect your business. Since you are a sociologist, I thought you would be interested in that aspect of why a venture capitalist like myself is interested in funding young Indian entrepreneurs.[10]

When asked why the founding members felt a need to create an organization specifically for South Asians, Dr. K. responded,

> Networking is an old practice in Silicon Valley. Everybody networks here. Indians needed their own entrepreneurial space. It is not as though we exclude any group. This is our [Indians'] version of the old boys' network. But we also have a larger interest in India—India's economy, to create an environment there for business and to improve Indo-US relations. That may not be of interest to other [ethnic] groups.[11]

As far as TiE and its members are concerned, the spread of free-market economics, to India and other regions in the Indian subcontinent, influencing policymakers and furthering ties between the U.S. and Indian economies, are entirely apolitical issues. "We are not in the business of politics, we are in the business of business," explained Dr. K when I pressed the issue further. Mr. S.C., president of the TiE chapter of the Carolinas, concurs with this view:

> I was part of the delegation of Indian entrepreneurs from the US to India. I think of my role as promoting trade relations between the two countries. For India, in the long run, this is an important issue. India has to develop free trade to a much greater degree, and Indians in the US can be the intermediary for this goal.[12]

However, the business of business is the business of politics as well. The transnational capitalist class of Indian entrepreneurs is, according to Sklair (1995), in the business of "taking on the political project of aligning their, and their co-nationals', interests in terms of the global capitalist system" (p. 134). The caveat I would add to Sklair's definition is that the reconceptualization of interests that is occurring among Indian immigrant entrepreneurs is not toward conationals (as both Dr. K. and S. C. are U.S. nationals). In the transnational business world of the software industry, Indian entrepreneurs in Silicon Valley and elsewhere in the United States are realigning their interests, and pushing forth a political project while simultaneously straddling two locations—their countries of origin and destination—as each is vital to their economic interest.

That TiE should carry political clout is not surprising. Asian Indians in the United States are a wealthy group, with the highest median household income of any racial category, including non-Hispanic Whites (U.S. Census, 2000). The networking organization for entrepreneurs is responsible for having created businesses with a total market value of $200 billion since the organization's inception in 1992 (TiE Web site). An exception to a rule made the importance of TiE even more emphatically clear to me. TiE meetings and conferences were very well attended by the local Indian entrepreneurial community in Silicon Valley, but it was not very common to see a non-Indian among the crowd. At one of the TiE meetings, a lone, young, white male told me that he had arrived as a guest of a TiE member, his business partner, an Indian professional in the Valley. He explained that he had deliberately decided to team up with an Indian so that he could find a way to network through TiE. "Being non-Indian, I was not sure how I would be received if I came here by myself, so I wanted an Indian business partner who would make it easier for me."[13] Even though this was not a frequent occurrence, the idea that a "native" would rely on an immigrant's social networks was entirely unexpected. I bring this piece of ethnographic data to the reader's attention to reiterate the enormous influence of transnational entrepreneurs in Silicon Valley.

TiE's interest in politics (even though the organization denies any such interest) is reflected in the general rise in Asian Indian political activity in the United States. Between the 1980 and 2000 elections, political contributions by Asian Indians have burgeoned from "almost nothing to approximately $8 million in a single election cycle" (Cho & Lad, 2004, p. 250). The authors also conclude that in terms of party affiliation, while Asian Indian groups show weaker attachment to the Democratic Party than Blacks and Latinos, their preferences lean in the direction of supporting Democratic candidates. However, the software community in both India and the United States watched the 2004 U.S. presidential election from the prism of the outsourcing issue, worried about Democratic candidate John Kerry's political rhetoric criticizing U.S. corporations that outsourced white-collar jobs to India. The victory of President Bush was received with jubilation within India's hi-tech business community. As India's stock index soared when the results were announced, Kiran Karnik, president of NASSCOM, India's software industry group, said, "We are very happy that Bush is back. The president's track record has been of recognizing the advantages of free trade" (quoted in Rai, 2004).

Thus, the transnational capitalist class of Indian entrepreneurs is engaging in a politics not from "below," resisting dominant structures, but together with national leaders and state bureaucrats, are engaging in political action in a transnational space between sending and receiving countries to promote the global capitalist system of free-market economics.

Indian Software Professionals: A "Valued Commodity"

The last of the points outlined by Sklair in identifying a transnational capitalist class, particularly in Third World settings, is the ability of this class, under certain circumstances, to dictate its own terms to foreign interests.

The circumstance that has provided some degree of control to Indian software professionals is the global demand for skilled IT workers. Although Germany refuses to be called an "immigration state" (Vertovec, 1996), an estimated 75,000 job vacancies in the IT industry forced a reshaping of the country's migration policies. In 1999, Germany announced a "green card" scheme that would admit 20,000 IT professionals from overseas locations to live and work there for a restricted period of 5 years. This measure was devised to prevent the permanent settlement of those who arrived under the new policy, as had happened with the guest workers in previous decades. To the surprise of the German government, 1 year after the pilot scheme was established, less than half the available visas had been claimed (Iredale, 2001; Meijering & van Hoven, 2003).

The scheme was reported widely in the daily newspapers in India, which was at a time when I was conducting fieldwork in Bangalore. Having seen an announcement in a local daily about a visa information session organized by the German consulate in the city, I decided to attend the meeting. The event was hosted in a banquet hall in a five-star hotel. The seating capacity was 300.

To the embarrassment of the organizers, an hour into the meeting, barely 30 potential "immigrants" had arrived. One of the attendees was a software professional employed by a reputed Bangalore-based firm. He explained his reasons for coming to the session and surmised as to why it was so poorly attended.

> I just came to see what it was about. I am not sure whether I want to go. Maybe for a short-term assignment. It's good to travel, see different places, get a different experience. Five years—that is too long. I think people like me would prefer to go to America, *if* they decide to leave India for good. I have a nice job here. You've heard of Wipro, right? It is one of the best IT companies in the world. Going to Germany would be like starting from scratch. But I was curious, that is why I came…Indians are not really interested in Germany because we know all the bad sides—it is too cold, we don't speak the language, did you see all the racist cartoons they have in their [German] newspapers about this green card scheme? After seeing that, why should we go? Maybe if you work for a smaller company in Bangalore, or maybe some other city and you don't get a chance to go somewhere else, I think those are the people who will use this scheme…People who work in Bangalore have a lot of different opportunities.[14]

The announcement by the German government of the new green card scheme for IT workers created a furor, with German politicians running campaign slogans such as *Kinder statt Inder* (Children instead of Indians) and German newspapers printing racist cartoons. The Indian media reported these incidents, not only within India but also in publications for the Indian community in the United States (Meijering & van Hoven, 2003), and in my conversations with those present at the visa information meeting, it was clear that the software professionals were well aware of the negative press against the new immigration policy offered by the German government.

The transnational software professionals are exercising their options in their choice of destination country, recognizing that they are a "valued commodity," as one interviewee put it. Despite the attractive package offered by the German government to these IT professionals, the power, it seemed, had shifted in the direction of the employees in the industry. In my interviews at prominent software houses in Bangalore, most expressed satisfaction with their quality of life in India and said that they would be motivated to leave for another country for greater travel and cultural experiences,[15] but not for a better quality of life. Having spent time on short-time assignments in various locations, the lure of becoming an immigrant was considerably muted. Thus, the interviewee quoted just previously, reflecting on his own choices, indicated that a short-term assignment would be preferable to the option of living in Germany for 5 years. Added to this sentiment was his pride in working for a "world class company," making the opportunity to move to Germany akin to "starting from scratch."

In a back-and-forth e-mail correspondence with an Indian software professional whom I had met at the visa information meeting and who had ultimately decided to apply for the German visa, the decision was explained to me thus: "I could not get an H-1B visa for the US this year since the quota had already been filled and I don't want to wait another year for a chance to go overseas."[16] His motivation was to "get foreign experience" so that he may be able to return to India in the near future and improve his career chances. Khadria (2004) also shows that highly skilled workers in India in the IT sector view their prospects in India to be attractive, but points to the fact that this is not true for other high-skilled professionals—nurses, for instance, do not consider their career opportunities in India to be fulfilling. Moreover, the author concludes from his study of IT professionals in Bangalore that "none of the professionals in Bangalore gave priority to the idea of settling down abroad" (Khadria, 2004b).

It is not just Germany that is facing shortage of IT workers and turning to India to fill these labor demands. The Australian government identified India as the most important source country for its projected need of 180,000 workers in 2004 (Xiang, 2001). Similarly, New Zealand also witnessed a growth in highly skilled immigrants, with India being the top sending country in 2000–2001 (Benson-Rea & Rawlinson, 2003). Britain too relaxed its work permit rules for IT specialists, a significant change in immigration policies of the last 30 years (Khadria, 2001).

The illustration using the German green card case is not to suggest that immigration flows are now entirely dictated by Indian IT workers or that country-level immigration policy or local labor demands are not a factor in creating global movement of labor. Nor is it the case that immigration of skilled workers from India to the United States has halted. To the contrary, it remains the case that most individuals who enter the United States under the temporary H-1B visa convert these to permanent residency status (Varma & Rogers, 2004). Furthermore, there are differences among IT workers based on level of education, skill, and experience. I am not, in other words, moving away from the argument that labor flows are still dictated by structural factors. However, brain-drain studies of skilled labor immigration singularly highlight the "push" from a developing country and the "pull" into a developed one. What skilled Indian IT workers are experiencing is a choice in where they might be "pulled," and in so doing, are dictating their own terms to foreign interests who have a desire to appeal to this population.

In the previous section, Sklair's usage of the term "transnational capitalist class" was applied to understanding the economic and political practices of Indian entrepreneurs and professionals in the software industry. I turn now to discuss the ideology behind transnationalism.[17]

Imagined Communities and (Trans)Nationalism

> Nation is an imagined political community. It is imagined because the members of even the smallest nation will never know most of their fellow-members, meet them, or even hear of them, yet in the minds of each lives the image of their communion.
>
> —Benedict Anderson (1995, p. 6)

> Nationalism is not the awakening of nations to self-consciousness: it invents nations where they do not exist.
>
> —Ernst Gellner (1964, p. 169)

As nations are imagined, so is (trans)nationalism. Anderson (1995) explored the processes through which the idea of the "nation" became a universally accepted form of knowledge. One of his main arguments was that nations came to be imagined with the rise of print capitalism—printing made possible the spread of "blueprints," "models," or "concepts" of nationalism to ever-increasing segments of the population. The printed word was used to create shared languages, histories, stories, and symbols that gave permanency to these imaginations. The idea of a nation does not exist in the memory of peoples or in their oral traditions. Instead, by the 2nd decade of the 19th century, after the historically decisive American and French revolutions, Anderson argued, "a 'model' of 'the' independent national state was available for pirating" (p. 81) through printed media. Importantly, these ideas were those of intellectuals, those in the ruling minority, since illiteracy was a common feature of many of the societies upon which the ideas were being thrust.

In the section that follows, I borrow from Anderson's toolkit and show how the imagined community of Indian transnationals, particularly that of the transnational capitalist class of Indian entrepreneurs, is being invented through the use of print media, the internationalization of professions, and scientific and managerial education.

Print Media

Arguing that the global capitalist system rests on the proposition that without a coordinated global economy there would be no transnational capitalist class, Sklair (2001) shows how the printing of the *Fortune Global 500* provides a way for the discourse of the "global" corporation to come into play. The list includes corporations from several different countries, and the underlying logic is that national contexts do not matter. By applying certain "universal" criteria—foreign direct investment, benchmarking, global corporate citizenship, and global vision—a list of 500 corporations can be attained. Similarly, Morgan (2001) points out the efficacy of print media in constructing the idea of the global corporation. He argues that a range of publications, including *The Financial Times, Wall Street Journal, The Economist, Fortune* and *Business Week*, deliver not only news but also commentaries on business practices, offering universal remedies and prescriptions.

I use these ideas to explore the ideas behind the magazine *Silicon India* and its own annual listing of top 100 companies owned and operated by Indians in the United States. Begun 7 years ago, this monthly magazine (there is also a web edition) would carry on its cover the subtitle, a "technology and business magazine." Now, in an era of increasing economic transnationalism between the sending and receiving countries, *Silicon India* changed the subheading 2 years ago to a "business and technology magazine in the U.S. & India." In keeping with its transnational outlook, the magazine has two offices—one in Silicon Valley and the other in Bangalore.

A founder of *Silicon India* explained the impetus behind the development of a magazine geared toward the Indian business community, particularly those in the hi-tech sector.

> There was a real hunger in the Indian business community in Silicon Valley to have our own monthly magazine. Our concept for this magazine was to focus just on hi-technology and business. Like, our own version of *Fortune* or the *Economist*, or maybe like *Business Week*. We thought that the mainstream press could actually use a magazine like ours that focuses just on business issues related to Indian-owned companies. There are topics we cover that would be of interest to Indian immigrants—visa issues, recruitment of IT workers, but overall, the idea was not to cater just to some specific Indian-only niche. We are not interested in carrying matrimonial columns, for instance. This is a business magazine, through and through. We are read not just by Indians but others as well...The advantage we have over other groups is that all Indians speak English, so we didn't have to worry about what language to use. We also started a web version so that the hi-tech business community, anywhere in the world, can access the content we provide. Given the global nature of the businesses run by Silicon Valley entrepreneurs, there is an interest in the business news coverage we offer that would be of interest to everybody, not just Indians.[18]

Silicon India, then, looks to the mainstream business press as its role model. The thrust of the magazine is to offer content that would be of appeal and interest to a wide audience. Like the mainstream business press, *Silicon India* carries lead stories on successful entrepreneurs, with a focus on the nature of their business and aspects that enable the success (several issues carry on the cover page the picture of a leading Indian immigrant entrepreneur), provides up-to-date information on business acquisitions, market capitalization, and initial public offerings by Indian-owned companies and opinion columns related to the global hi-tech industry. The magazine provides publicity to Indian entrepreneurs and professionals through features that include "CIO of the Month," CIO referring to Chief Information Officer, and "Start Up of the Month." There is also coverage given to the philanthropic activities of Indian entrepreneurs, in the style of global citizenship issues in the mainstream business press, and the online version provides a list of United States-based nonprofits that have ongoing projects in India.

It is true that the magazine is geared toward the business community of a specific ethnic community, but the contents are devoid of any topics that are uniquely ethnic, such as the printing of matrimonial columns. These matrimonial columns, a staple of other Indian magazines, including *India Abroad* (printed in the United States) or *India Today* (based in India but with a separate U.S. print version), would shift the focus away from the idea of providing content that would appeal to a larger audience. Equally importantly, the magazine is in the *lingua franca* of the hi-tech world, a point brought out by the respondent. The use of English is another reason why the magazine hopes not to relegate its readership to an ethnic-only community. The topics covered—immigration, outsourcing—would potentially be of interest to other non-Indian companies either seeking global labor or looking to India as an outsourcing destination. Further, as Vertovec (2002) points out, *Silicon India*'s online edition provides a network for information exchange and recruitment among occupational professionals, and through its extensive directory of companies, adds another layer by offering information for those individuals and companies seeking to outsource their IT functions to India. The online version serves another function: it creates an imagined community of transnationals, which is not specific to this magazine, but speaks to the general nature and power of the Internet. Anderson (1995) claimed that the concept of the nation could arise "only when substantial groups of people were in a position to think of themselves as living lives parallel to those of other substantial groups of people—if never meeting, yet certainly proceeding along the same trajectory" (p. 188). The shattering of time via the Internet allows people in various geographical locations, operating in different time zones, to access the information simultaneously. The reach of the Internet is an important part of the idea of creating a shared notion of "globality," and those who have access to such media are able to conceive of themselves as "transnational" or "global" citizens.

Following in the footsteps of *Fortune 500*, *Silicon India* also provides its own yearly list of top companies, in what it calls the si100. The list provides the name of the founder, year it was founded, investors, ownership (private or public, and, if public, the Nasdaq listing), customers, revenue, and a description of business activity. The list has come to formalize the status of hi-tech Indian-owned companies in the United States, with many interviewees telling me with some pride that their company was on the "si100 list." Not even the "dot-com bust" of the early 2000s dampened the efforts. In fact, the 2001 list carried the following statement: "What a great time to test every entrepreneur, professional and investor's real merit, drive, ideas and commitment, and separate the men from the boys" (September 2001, p. 24). That year's list did not carry any companies in the consumer Internet space, arguing, "most Internet and e-commerce companies are as much hi-tech companies as your neighborhood convenience store." The list carries with it profiles of specific companies, and many of them are listed as being one of *Fortune 1000* or *Global 2000* companies. Not all companies have India operations, but where they do, these are highlighted in the company profiles.

Thus, *Silicon India* is one of the tools used by the increasingly global elite of Indian immigrant entrepreneurs in the hi-tech industry, to create an imagined community of transnationals. It allows the transnational capitalist class to use a common language, English, but also a common language of the business world, one which views the world as being increasingly globally interconnected, or, at least, transnationally connected. The profiles of top Indian businessmen provides a vocabulary and a "blueprint" through which to share common stories, create myths around successful entrepreneurs, and establish a "mode" ' for conducting socially responsible business for fledgling entrepreneurs. Catering to the Indian technology community in India and the United States, the publication also seeks to generate wider interest in mainstream society, and as such, distances itself from being considered a purely "ethnic" business magazine.

Internationalization of Higher Education: Shared Languages

Indians in Silicon Valley were able to ease into entrepreneurship activities in part because of their advantage as English speakers, but also as a result of the language of science and technical education. The brain drain was possible because of the perceived similarities in higher education in the science and engineering fields. An Indian professor of engineering explained it to me thus:

> Every discipline has its own jargon. I arrived here [the United States] with my wife. I came as a graduate student in engineering. She came as a graduate student in psychology. I had no problem with my grad level courses. They were a breeze after my training at IIT. It was all so easy. What they were teaching me at the graduate level is something I had already done as an undergraduate back home [in India]. My wife had a different experience. I think she had to learn a lot of technical jargon she was unfamiliar with from her training in India.[19]

Founded in the likeness of U.S. universities such as MIT and Caltech (see chapter 2), the Indian Institutes of Technology were able to provide comparable education, with English as the medium of instruction and textbooks that were known worldwide in their respective disciplines. Students were also exposed more intimately to Western pedagogy through guest faculty from overseas, primarily U.S. universities. By contrasting his experience and that of his wife's in a social science discipline, the respondent is bringing attention to the way education in the sciences has been able to generate a shared arena of knowledge that cuts national boundaries, thereby easing the incorporation of immigrants into their newly adopted societies. However, the quality of training provided by Indian institutes of higher learning could not overcome issues of accreditation. Thus, many future workers in the U.S. high-skilled industries first arrived as graduate students—in my own research sample, of the 26 entrepreneur interviewees in Silicon Valley, all but 1 had received training at U.S. universities. The impact of the training was highlighted by yet another Indian immigrant interviewee, who contrasted his experience as a graduate student in engineering with that of other international students. "The difference betweens Indians and other international students was that we could do the math and so could they but we could also explain the problem in English, something others struggled with."[20]

Morgan (2001) points out yet another way education helps the production of an imagined community of transnationals and creates a shared language through standardized curricula: the MBA degree. He argues that most top MBA schools aim to produce a "general manager," and these programs are conscious of the identity-forming functions of the process. Indian immigrants into the United States who arrive as graduate students are overwhelmingly in the fields of science and engineering, yet there has been a steady rise of students entering MBA programs, and oftentimes, obtaining a combination of the two degrees (Khadria, 2004b). It is perhaps not surprising that the Managing Director of one of the most prominent global consulting companies, McKinsey, is an Indian immigrant, Rajat Gupta, who, incidentally, is also a graduate of the Indian Institute of Technology. The shift in the occupational categories in the period of brain drain and the present period where entrepreneurship is on the rise can also be seen through the lens of the rise of Indians in management positions and the effect of Indian MBA programs rising to the same rank as top engineering schools. According to an immigrant entrepreneur in Silicon Valley,

> Everybody talks of IIT, but not many people realize that IIM [Indian Institute of Management] has an equally important status. IIMs have a lot of cache in Silicon Valley because they teach us like any other MBA school in the States. IIT can train you to be a good engineer but these management schools train you to be good entrepreneurs,

> good managers. You learn an approach to managerial problem solving, not just how to fix a bug in your computer. You can do different things than just with an engineering degree.[21]

Thus, as Morgan pointed out, "[MBA programs] and consultancy companies...bring packaged solutions to business problems across the world. These solutions may be more or less customized and innovative depending on the cases and problems but they remain built on the notion of general management skills as propounded in the Anglo-American framework" (p. 126).

In the current period of globalization of the hi-tech industry, yet another significant shift has occurred. Certain professions, particularly in the field of IT, have been internationalized. This internationalization of professions implies a convergence toward international standards and procedures away from nationally defined standards and national forms of regulation (Lenn & Campos, 1996). This is yet another way that a shared language, and therefore, a shared ideology of transnationalism are generated. Iredale (2001) lists a number of characteristics of the IT industry that has allowed it to move out of the arena of national controls: domination by the English language as the basis, accreditation of qualifications driven by private companies, on-the-job experience as the most important means of acquiring human capital or becoming multiskilled, a high level of intra- and intercompany and interregion/country mobility, and a profession that is largely unregulated by unions or other mechanisms.

The issue of accreditation of qualifications of large corporations has proved critical in propelling Indian immigrants to the forefront of the globalization of skilled workers, and allowed the transnational business activities of Indian entrepreneurs, both in Silicon Valley and within India. Companies such as Microsoft and Oracle have their own certification programs, and Indian training institutes are well versed in providing these skills at a far less cost than comparable technical institutes in advanced countries. Free of national controls, young men and women trained in these specific skills find themselves very well placed in the competitive arena of information-technology jobs.

The standardization of skills allows Indian IT workers to become transnational workers, traveling constantly in and out of countries to provide software solutions across much of the developed world. In one instance at a cafeteria at the headquarters of a local Bangalore software company, I did an informal count in a group of 10 employees. Each had been to at least two separate overseas locations, and two counted as many as five. The ease of movement of workers testifies to the ease of movement of work.

In conclusion, what has been argued in this section is that education, whether in the fields of science and engineering or through the MBA program, provides a shared body of knowledge that can be called upon to provide a shared language of global business solutions. In the same way, the growth of the internationalization of IT skills ensures that workers who possess these skills begin to view themselves as transnational citizens of the world.

CONCLUSION

This chapter has traversed a long path. It began with an argument put forth in the existing body of work on economic transnational activities of immigrants—that the entrepreneurial practices of transnational immigrants should be viewed as a form of transnationalism from "below." It then posed the question for a specific immigrant group, namely Indians, and asked whether the practices of highly skilled immigrants mimic those of immigrants arriving with disadvantages in the labor market. I have argued that Indian immigrants in the hi-tech industry view their transnational businesses as a means of pushing forward an idea of transnationalism that is more akin to how dominant groups view their global practices. Thus, for Indian entrepreneurs, both in Silicon Valley and within India, it is hard to make the case that their transnationalism is a globalization from "below."

ENDNOTES

1. The issue of whether low-skilled migration is demand driven or spurred by a supply of people eager and willing to migrate has generated considerable academic debate (see Staring, 2000, for a discussion). For a social network approach—the idea that migration between sending and receiving countries becomes self-sustaining because of the existence of dense social networks—see Light, Kim, and Hum (2000) and Burgers (2000).
2. Castells and Portes define the informal economy as all income-earning activities that are not regulated by the state in social environments where similar activities are regulated (Castells & Portes, 1989).
3. I use Dicken's definition of transnational corporation. " 'Transnational corporation' [is preferable] to the more widely used term 'multinational corporation,' simply because it is a more general, less restrictive term. The term 'multinational corporation' suggests operations in a substantial number of countries whereas 'transnational corporation' simply implies operations in at least two countries, including the firm's home country. In effect, all multinational corporations are transnational corporations but not all transnational corporations are multinational corporations. More simply transnational corporations—or TNCs—provide 'the means of co-ordinating production from one centre of strategic decision making when this co-ordination takes a firm across national boundaries' " (Dicken, 2003, p. 21).
4. Sklair (2001) distinguishes the transnational capitalist class in Third World countries in the present global age from the *comprador* class during the period when most of the third-world was colonized by Western powers. During colonial times, the *comprador* class aligned itself with specific foreign countries, for example, the Indians with the ruling British power, serving the interests of the foreign power over and above the interests of their own conationals.
5. Personal interview with Rajeev M., entrepreneur in Silicon Valley, October 1999.

6. Personal interview with Subroto B., entrepreneur in Bangalore, May 2000.
7. Personal conversation with S. C, founder of TiE in the Carolinas, January 2004. S.C. was one of the TiE members who had accompanied President Clinton on his India trip. See also, Nolan (2001).
8. Indus is a reference to the ancient Indus Valley Civilization on what is now the Indian subcontinent.
9. Personal interview with Dr. K, Silicon Valley entrepreneur, November 1999.
10. Personal conversation with Dr. Matthew T, Silicon Valley venture capitalist, TiE meeting, November 1999.
11. See endnote 9.
12. Personal conversation with S. C, founder of TiE Carolinas, in January 2004.
13. Personal conversation with Chris S., at a TiE meeting, in January 2000.
14. Personal conversation with Siddharth S., software professional in Bangalore, March 2000.
15. The motivation for greater cultural experiences is corroborated by an ethnographic study of Indian IT professionals in Germany (see Meijering & van Hoven, 2003).
16. Email correspondence with Ravi V., software professional in Bangalore, May 2000.
17. The discussion is based on Morgan (2001).
18. Personal conversation with Yogesh S., January 2000.
19. Personal conversation with Dr. S, professor of engineering, October 2000.
20. Personal conversation with Suresh G., software professional in Silicon Valley, November 1999.
21. Personal interview with Raghu T., Silicon Valley entrepreneur, December 1999.

Conclusion

Like the product cycle of technological innovations, any treatise on the software industry runs the risk of quickly becoming obsolete. The shift in the immigration pattern of highly skilled workers from India to the United States—from brain drain to the bodyshopping of temporary workers—occurred over a long period of 3 decades. Yet the change, from bodyshopping of labor to outsourcing of business processes—that is, the shift in the movement of people to the movement of jobs—happened within a comparatively quick span of half a decade. To predict what the next cycle will entail is difficult. Thus, this study has taken the approach of *la longue duree.*

The book has explored the roots of India's highly valued scientific and technological personnel and traced the roots back to the science and technology policies of British colonial rule and India's own postcolonial embrace of scientific socialism. From there, I have connected India's educational policies to the labor-force needs of the United States that resulted in the significant presence of Indian immigrants in Silicon Valley. Highlighting how the unique techno-entrepreneurial culture of the Valley made its imprint on the "ethnic" entrepreneurship of its Indian immigrant populace, I have argued that the business practices of Indians must be understood within a context of transnationalism. Finally, the growing confidence of Indians in the software industry—its entrepreneurs, professionals and venture capitalists, both in the diaspora and in India—shows that this new group of middle-class Indians is more akin to a transnational capitalist class aligning with the global capitalist system and less the product of a developing country who are mere pawns caught up in the forces of globalization.

In an increasingly globalized world of the 21st century, the growing presence of Indian immigrant entrepreneurs in Silicon Valley, the connections they are establishing with their home country, the rapid expansion of the Indian software industry, and the rise of a new, wealthy group of middle class Indians have implications for both the United States and India. Thus, I conclude with some remarks on pertinent issues of public policy for these two countries, and following from these, the potential for conducting future research.

POLICY CONSIDERATIONS

The United States has sought to meet the challenges of international competitiveness through off-shoring labor-intensive segments of industries and retaining the technological and marketing aspects of the production process. This is as true today of traditional industries such as textiles and apparel as it is of knowledge-based industries like information technology. The process of industrial restructuring in the textiles industry has meant that while outsourcing has denuded the American labor force of well-paid blue-collar jobs, it has at the same time allowed manufacturers to remain competitive by producing technologically innovative products. Labor, in other words, is being substituted by technology (Siekman, 2004).

The expectation remains high that the information-technology industry in the United States will remain more competitive globally due to the current phenomenon of outsourcing of white-collar jobs. The argument in favor of outsourcing states that technology companies in the United States will increase productivity and boost efficiency by focusing on their core competencies—developing the latest innovative products and allowing labor force in other countries to manufacture, test, and mass produce them. This is the logic that prompted IBM to sell its unprofitable PC business to the Chinese PC manufacturer, Lenovo, in December 2004 (Cringley, 2004). Moreover, some claim that the impact of outsourcing on loss of American jobs is exaggerated (Drezner, 2004). In a similar vein, in two separate studies, Slaughter has argued, firstly, that U.S. multinationals have as much been a source of job creation as job destruction (2004a), and secondly, that the debate on the globalization of jobs should not only consider the outsourcing of jobs by American firms to overseas locations but also focus on the "insourcing" of jobs into the United States by foreign-headquartered companies (2004b).

While taking a stance on the benefits and pitfalls of outsourcing is out of the scope of this work, the data presented allow us to add an element to the debate, that is, the influence of immigration of highly skilled workers. The United States has relied on a foreign labor force whenever shortages arose within the domestic labor market. This was also the case within the software industry, with Indians arriving in large numbers in the 1990s to fill labor gaps. It is now well known within academic as well policy circles that Silicon Valley's foreign personnel not only make up a significant part of the region's labor pool, but also that a fourth of new companies founded in the Valley are by first-generation immigrants, primarily of Indian, Chinese, or Taiwanese descent (Saxenian, 2002). Add to this the little-known fact that it was an Indian professional working in Silicon Valley, Srikant Nadhamuni, who helped develop the Intel Pentium Chip, or that it was another Indian, Sabeer Bhatia, who founded Hotmail, or a third Indian, Vinod Khosla, who helped establish Sun Microsystems. It is clear from both survey and observational accounts that immigrants are crucial to the continued competitive edge of innovative hubs such as Silicon Valley.

The reliance on foreign personnel has worked to the advantage of Silicon Valley and to American industries in general. However, three factors collide to create a cautionary tale for the future. First, there is increasingly a process of return migration of many of the Valley's immigrant professionals and entrepreneurs. While the process of Indian return migration is less well established and relatively recent, the rates of Chinese and (especially) Taiwanese return migration have been documented in the literature (Saxenian, 2002).

Secondly, even though emigration from India continues to occur, either in the form of student immigration or the migration of skilled labor, there is increasing evidence of temporary migration for short-term assignments rather than permanent migration. Some of the migratory "push" toward the United States is tempered by the changes, whether real or perceived, of immigration laws after 9/11, making potential Indian immigrants wary of making the long migration journey to the United States (Mooney & Neelakantan, 2004). In general, the 2003–2004 academic year saw an overall drop in international student enrollments in U.S. universities, the first absolute decline since academic year 1971–1972 (Open Doors, 2004). Meanwhile, other countries, primarily Australia, Canada, New Zealand, and the United Kingdom, in response to labor shortages within their national borders, have begun to offer attractive immigration packages to potential high-skilled immigrants, thereby proving to be alternative and viable destinations for Indian professionals and students alike (Khadria, 2004a).

Third, the backdrop to the changes in immigration patterns of highly skilled labor, particularly in the areas of science and technology, is the issue of the ability of the domestic labor pool to sufficiently fill any labor shortages. An OECD report indicates that U.S. high school students rank below average in science and mathematics compared to similar-age students in other industrialized countries. One of the critical findings of the survey is that American students lag behind not only in generalized math scores but also in applying math skills to real-life problem solving, skills that would be of particular interest to future employers (PISA, 2003).

These seemingly unrelated events are important for the continued competitiveness of U.S. hi-tech industries and to prepare future generations for the challenges of the increasingly competitive 21st century workplace. Human capital is essential for technology firms that wish to focus on the most creative and innovative stages of the production cycle. U.S. institutions of higher education can nary afford to lose foreign students, as traditionally, this “semi-finished human capital,” upon completing their degrees in the United States, typically enters the permanent labor market in the country. Approximately one-fifth of all the doctoral degrees awarded by U.S. institutions and one-third of doctorates in engineering, mathematics, and the physical and biological sciences are earned by international students. A large segment of the entrepreneurs of Indian origin in Silicon Valley first arrived as graduate students. If the reliance on immigrant labor is in jeopardy, due to stricter immigration laws or because the option of return is increasingly possible, then domestic educational policymakers must focus on improving science and math skills among the local student body.

On the other hand, the change in the brain drain of labor to the transnational mobility of highly skilled personnel has been of considerable advantage to the Indian software industry. The role of diaspora communities in fostering foreign direct investment in origin countries is gaining increasing academic attention (Ramamurti, 2004). The Indian government too has taken note of the growing strength of the Indian entrepreneurial community in the United States, and has begun to offer dual citizenship for persons of Indian origin, which does not allow voting rights but enables the diaspora community to invest economically in India.

India is reaping the benefits of its post-Independence policies of higher education promoting science and technological training. The qualified personnel produced by the nation's schools of engineering and sciences have played a significant role in the phenomenal growth of India's software industry, as they have in the economy of Silicon Valley and other technological hubs across the United States. According to an OECD report, in the year 2000, the total stock of engineering personnel in the country was almost 2.5 million (Khadria, 2004a). The large labor pool should continue to meet India's labor needs in the foreseeable future. The Indian IT sector has proved to be the country's fastest-growing segment—even in troubled times, during the challenging economic environment of 2001–2003. The Indian software industry has grown from a mere U.S.$150 million in 1991–1992 to $16 billion by 2002–2003 (NASSCOM, 2004b).

While these figures attest to how the past decade has proved to be extremely fruitful for the Indian software industry, from a larger economic and social development viewpoint, two issues need to be considered. Firstly, even though India is highly competitive in a knowledge-based, hi-tech industry, large sections of the country's population remain unable to directly benefit from the growth of the IT industry, not least because India presents the observer with an incredible anomaly. It produces the world's highest number of engineering personnel, but also has the most number of illiterates in the world (Chopra et al., 2004). Over the past decade, the middle class in India has expanded, creating a huge market for consumer goods. However, in the absence of benefits that accrue to ever-widening segments of the population, India presents a scenario of creating "developed" islands in a sea of "underdevelopment."

A second issue is that Indian IT companies have matured in the last decade, providing more value-added services to customers. In the early 1990s, companies in India were in the business of providing services in the least creative segments—those of routine coding and testing. The Y2K issue allowed the entry of Indian IT companies into the U.S. market and, more importantly, for American companies to develop successful business relationships with Indian companies and gain trust and confidence. Within the span of a decade, software companies in India moved from merely providing the cheapest labor in the least innovative stage to increasingly performing well in higher value-added segments, including original research and development. However, in the same way that U.S. technology companies have to restructure in order to retain the most innovative segments of the production chain, similarly, Indian companies have to foresee how they might be able to enter a hitherto uncharted terrain—that of creating products that they are able to market themselves.

If there is any lesson to be learned from the brief history of late capitalist development, it is that countries or industries can no longer rely on competence in any specific segment of the production cycle—in order to remain competitive, companies have to constantly climb the long and inexorable ladder to the top of the innovation cycle. How India will perform in making the shift from being primarily an exporter of software services to a producer of software products will determine the longevity of the industry's viability in the country.

DIRECTIONS FOR FUTURE RESEARCH

Based on the issues raised in the previous text, a number of interesting questions arise that would require further and careful research attention. The first, and perhaps most obvious, is the extension of the research conducted on Silicon Valley's Indian immigrant population to other similar technology hubs within the United States. While the Route 128 area in Massachusetts has been the subject of study (Saxenian, 1994), the issue of the region's immigrant population is yet to be untangled. Furthermore, according to a UNDP report, the Raleigh-Durham-Chapel Hill area in North Carolina ranks fifth in global technological hubs, behind Silicon Valley, Boston, Stockholm, and Israel (UNDP, 2001). Interestingly, the state of North Carolina has also witnessed increased immigration from India, which now ranks as the second-largest sending country to the state, behind Mexico (FAIR, 2003).

What makes such a study particularly interesting is that North Carolina, unlike California or the Boston area, is not a traditional immigrant-receiving state. The difference in numbers between, for example, Mexican and Indian immigrants is vast, however, as pointed out by Koser and Salt (1997), it is the case that migration of highly skilled workers oftentimes tends to be overlooked and becomes invisible because their small numbers easily render them statistically invisible. It is also because such people are not perceived to be a "problem," and their generally middle-class characteristics render them socially invisible. Whether the presence of Indian immigrants is critical to the growth of North Carolina's technology industry is a question that has yet to be answered and one that would be of particular interest to state policymakers.

A second line of research that would add to the debate on high-skilled immigration as well as on issues of competitiveness of U.S. companies is an examination of the composition of immigrant personnel within American technology corporations. What this study has shown is that the rise of entrepreneurship among Indian immigrants in one of the mainstays of America's economy, its hi-tech industry, and in one of the most highly valued regions of the country, Silicon Valley. It is imperative now to understand the extent to which top U.S. companies (non-Indian owned) are reliant on immigrant labor, especially in the crucial segments of research and development. Unraveling the human capital component of corporations is important in the debates on immigration policy to ensure the continued growth of innovation within U.S. companies.

Following from this is an issue that has relevance to the literature on the sociology of work. The 21st century workplace is very different from the 20th century workplace. As Kalleberg argues, part-time, temporary, and contract work are becoming commonplace in the organization of work arrangements (2000). As indicated in chapter 3, American companies are increasingly relying on the L-1 visa category that allows the temporary presence of high-skilled personnel within the United States. The use of this specific visa category allows companies to be flexible because they bring in personnel to work onsite as and when the demand arises. Secondly, the visa does not require American firms to pay the prevailing American wage rates, thus increasing profitability for companies. An analysis of the extent to which corporations rely on this visa would shed light on the changes in the American, and thereby, the global workplace.

Appendix A

List of Stanford University Electrical Engineering Professors With Industry Links

Stanford University Professors	Industry Links
John L. Hennessy President of Stanford University Professor of Electrical Engineering and Computer Science	Founded MIPS Computer Systems, now known as MIPS Technologies
Nick Bambos Professor of Electrical Engineering and Management Science and Engineering	Founding Director, Stanford Networking Research Center (SNRC), a major research and technology-development partnership between Stanford and the information-technology industry in Silicon Valley and beyond.
John M. Cioffi Hitachi America Professor of Engineering	Chairman, ASSIA, Inc. On the Board of ClariPhy Communications Inc., Teranetics Communications
Donald Cox Harald Trapp Friis Professor of Electrical Engineering	Before joining the Stanford faculty in 1993, he was the Executive Director at Bellcore and Bell Labs.
Robert Dutton Robert and Barbara Kleist Professor of Engineering	Cofounder of Technology Modeling Associates (TMA), now a division of Avant!
Abbas El Gamal Professor of Electrical Engineering	Cofounder of Silicon Architects, which is currently part of Synopsys
Dawson Engler Professor of Electrical Engineering	Cofounder of Coverity.com

(*continued on next page*)

Appendix A. (*continued*)

Stanford University Professors	Industry Links
Hector Garcia-Molina Leonard Bosack and Sandra Lerner Professor of Electrical Engineering and Computer Sciences	On the Technical Advisory Board of DoCoMo Labs, USA, Yahoo Search & Marketplace Venture Advisor for Diamondhead Ventures Member of the Board of Directors of Oracle and Kintera
James Gibbons Reid Weaver Dennis Professor of Electrical Engineering Special Counsel to the President for Industry Relations	Founded SERA Learning Technologies
Bernd Girod Professor of Electrical Engineering	Founder or Advisor of Polycom, Vivo Software, and RealNetworks
Mark Horowitz Professor of Electrical Engineering and Computer Science	Helped start Rambus Inc.
Gregory T. A. Kovacs Associate Professor of Electrical Engineering	Cofounder, Cepheid
Benjamin Van Roy Associate Professor of Management Science and Engineering and Electrical Engineering	Advisor for several technology companies
Jennifer Widom Professor of Electrical Engineering	On the Advisory Board of Abrevity and Ingrian Networks and consultant for StreamBase
Bruce Wooley Robert L. and Audrey S. Hancock Professor of Electrical Engineering	On the Board of integrated Silicon Solution Inc. (ISSI)

Appendix B

Selected List of Alumni of the Indian Institute of Technology (IIT) and Their Industry Affiliations

Names	Position	Company
Delip Andra	Founder & CEO	Minekey Inc.
Jiten Apte	Cofounder & CEO	Wizarth
Ravi Apte	Previously CTO	American Stock Exchange
Deepak Bhagat	Former Director of Engineering	Sun Microsystems
Abhay Bhushan	Chairman	ASquare
Dr. Pallab Chatterjee	Interim CEO	i2 Technologies
Dr. Purnendu Chatterjee	Chairman	Chatterjee Group
Jassi Chadha	President & CEO	marketRX
Sumir Chadha	Managing Director	Sequoia Capital India
	Chairman	US-India Venture Capital Association
Ashish Chona	CEO & Cofounder	InSync
Bharat Desai	Chairman & CEO	Syntel
Gururaj Deshpande	Chairman	Sycamore Networks
Rono Dutta	Former President	United Airlines
	Present Chairman	Air Sahara
Girish Gaitonde	Founder & CEO	Xoriant Solutions
Sunil Gaitonde	CEO & Founder	GS Lab

(*continued on next page*)

Appendix B. (*continued*)

Names	Position	Company
Rakesh Gangwal	Former President	US Airways
Ajei Gopal	Executive Vice President	Cendura
Ashwini Gupta	Executive Vice President and Chief Risk Officer	American Express
Dev Gupta	Chairman	Narad Networks
Dr. Naren Gupta	Cofounder	Integrated Systems Inc.
Rajat Gupta	Former Managing Director	McKinsey & Co
Umang Gupta	Chairman & CEO	Keynote Systems
Vinod Gupta	Chairman	InfoUSA
Venky Harinarayan	Cofounder	Jungle.com
Ashok Jain	Chairman & CEO	Teraburst Networks
Suhas Kakde	President & CEO	US Aeroteam
Hemant Kanakia	Partner	Columbia Capital
	Cofounder	Photuris
Nat Kaushik	Founder & CEO	Fineground Networks (acquired by Cisco)
Rajeev Khanolkar	Former CEO	Net Forensics
Vinod Khosla	Cofounder	Sun Microsystems
	Partner	Kleiner Perkins Caufield & Byers
Rajeev Laroia	Founder	Flarion Technologies
Arjun Malhotra	Chairman & CEO	Headstrong
Vivek Ranadive	Chairman & CEO	Tibco
Rakesh Mathur	Cofounder	Junglee.com
Shailesh Mehta	President & CEO	Granite Hill Capital Ventures
Victor Menezes	Senior Vice-Chairman	Citigroup
N. R. Narayana Murthy	Chairman	Infosys Technologies

(*continued on next page*)

Appendix B. *(continued)*

Names	Position	Company
Avi Nash	Director	Sigma Aldrich Corporation
Arun Netravali	Former President	Bell Laboratories
Pavan Nigam	Serial Entrepreneur; Cofounder	Healtheon/WebMD
Nandan Nilekani	President & Managing Director	Infosys Technologies
Bala Parthasarthy	Cofounder	Snapfish
Suhas Patil	Cofounder	Cirrus Logic
Narendra Patni	Chairman & CEO	Patni
Vasant Prabhu	Executive VP & CFO	Starwood Hotels & Resorts
Kanwal Rekhi	Cofounder	Excelan
Ashutosh Roy	Cofounder	eGain
Arun Sarin	CEO	Vodafone
Dr. Amarpreet Sawhney	President & CEO	Confluent Surgical Inc.
Schwark Satyavolu	CTO	Yodlee.com
Sudhakar Shenoy	Founder	Information Management Consultants
Pradeep Sindhu	Founder	Juniper Networks
Balraj Singh	Founder	Peribit
Gunjan Sinha	Founder	eGain and Viman Software Inc.
Prabhakant Sinha	Managing Director	ZS Associates
Noorali Sonawala	Founder & CEO	Sunrise Systems
S. Sunderesh	President & CEO	Candera
Ashish Vengsarkar	Founder	Photuris
Ramana Venkata	Founder	Stratify
Sunil Wadhwani	Cofounder	iGate
Dr. Romesh Wadhwani	Chairman & President	Wadhwani Foundation
	Founder	Symphony Technology
Prem Watsa	Former Chairman & CEO	Fairfax Financial Holdings

Appendix C

Immigrants Admitted to the United States From the Top Five Countries of Last Residence, 1827–1997

1821–1840	1841–1860	1861–1880	1881–1900	1901–1920	1921–1940	1941–1960	1961–1980	1981–1997
Ireland	Ireland	Germany	Germany	Italy	Canada	Germany	Mexico	Mexico
Germany	Germany	U.K.	U.K.	Austria/ Hungary	Germany	Canada	Canada	Philippines
U.K.	U.K.	Ireland	Ireland	Soviet Union	Italy	Mexico	Cuba	China
Other	France	Canada	Italy	Canada	Mexico	U.K.	Philippines	Dominican Republic
	Canada	Norway/ Sweden	Austria/ Hungary	U.K.	U.K.	Italy	U.K.	India

Source. *Statistical Yearbook of the Immigration and Naturalization Service*, U.S. Department of Justice, 1998.
Note. Countries listed from top to bottom of a column represent descending order of sending countries to the United States.

Appendix D

Emigration From India to the United Kingdom and the United States, 1964–1990 (Selected Years)

Year	Number of Persons	
	United Kingdom	United States
1964	13,000	634
1965	17,100	582
1966	16,700	2,458
1969	11,000	5,963
1970	7,200	10,114
1971	6,900	14,310
1978	9,890	20,753
1984	5,140	24,964
1986	4,210	26,227
1988	5,020	26,268
1989	4,580	31,175
1990	5,040	30,667

Source. Khadria (1999, p. 62).

APPENDIX E

LEADING EMPLOYERS OF SPECIALTY OCCUPATION WORKERS (H-1B), OCTOBER 1999 TO FEBRUARY 2000

Rank	Company	Number of Approved Petitions
1	Motorola Inc.	618
2.	Oracle Corp.	455
3.	Cisco Systems Inc.	398
4.	Mastech (now iGate)	389
5.	Intel Corp	367
6.	Microsoft Corp	362
7.	Rapidigm	357
8.	Syntel Inc.	337
9.	Wipro Consultancy Ltd	327
10.	Tata Consultancy Services	320
11.	PriceWaterhouseCooper LLC	272
12.	People Com Consultants Inc	261
13.	Lucent Technologies	255
14.	Infosys Technologies Ltd	239
15.	Nortel Networks Inc	234
16.	Tekedge Corp (now Xoriant)	219
17.	Data Conversion	195
18.	Tata Infotech	185
19.	Cotelligent USA Inc	183
20.	Sun Microsystems Inc	182

Source. Leading Employers of Specialty Occupation Workers (H-1B), U.S. Immigration and Naturalization Service, June 2000.

Appendix F

Summary Statistics of Interviewees

A. Breakdown by Position

Silicon Valley

Entrepreneurs of Indian origin	26
Employees in managerial positions in Indian-owned firms	13
Indian executives in non-Indian-owned firms	6
Indian software professionals (employees not in managerial positions)	13
Total	58

Bangalore

Entrepreneurs	14
Executives (managerial position)	9
Software professionals	5
Total	28

B. Gender Breakdown of Interviewees

Silicon Valley

	Male	Female
Entrepreneurs	24	2
Employees in managerial positions (Indian-owned firms)	12	1
Indian executives	6	0
Indian software professionals (employees not in managerial positions)	6	7
Total	48	10

(*continued on next page*)

Appendix F. *(continued)*

Bangalore

	Male	Female
Entrepreneurs	14	0
Executives	7	2
Software professionals	1	4
Total	22	6

C. Return Migrants in Bangalore

Entrepreneurs[1]	4
Executives	1
Software professionals[2]	0

Endnotes

1. Of the four return migrants, three had returned from the United States and one from Sweden.
2. I am using the term "return migrant" for those who had immigrated to another country and then made the decision to return to India. All the software professionals in the sample had spent time overseas on work assignments, although none had made a decision to immigrate for good.

Appendix G

Interview Schedule

Name____
Age____
Male Female
Name of Company_____________________________
Title/Position________________________________
Citizenship/Visa Status__________________________
Marital Status________________________________

Questions for Indian Entrepreneurs in Silicon Valley

A. Immigrant History and Educational and Professional Background

1. When did you immigrate to the United States?__________
2. In what capacity?
   - (a) _______Student (F1 visa)
   - (b) _______With a job (H-1B or L-1) ________(note which one)
   - (c) _______With an immigrant visa (Green Card)
   - (d) _______Other____________________________
3. How old were you when you immigrated to the United States?
4. What motivated you to come to the United States?
5. Where in India are you from?

6. What is your educational background? (Write down degrees obtained and names of institutions).
7. What job(s) did your father and mother have?
8. How would you describe the family you grew up in?

   (a) Lower class____
   (b) Middle class_______________________________________
   (c) Upper class____
   (Note separately if the answer is lower middle, upper middle, etc.)

9. What is your professional background? (Write down name(s) of companies and position(s) held.)

B. Entrepreneurial History

1. When did you start your company? Alone, or in partnership?
2. Why did you decide to start your company?
3. Why did you choose to start your company in Silicon Valley?
4. How did you get funding to start your company?
5. Were any Indian associations, groups, or individuals important in your setting up the business? Do they continue to be important? If yes, explain. If not, why not? (Do you attend TiE meetings? If so, why? If not, why not?)
6. What are the challenges you have faced in running your business?

C. Nature of Business

1. Does your company have any ties to the Indian software industry? If so, what? How important are these ties?
2. Is your education and work experience related to the business you are running now? How would you evaluate the impact of your educational/professional training in the running of your business?

D. Open-Ended Questions

1. What does your family think about your entrepreneurship activities?
2. How important has your family been to your running the business?
3. Why do you think there is such a strong presence of Indians in Silicon Valley?
4. Is it difficult to be an immigrant and run your own business in America?[1]
5. What, according to you, are the key elements in running a successful company?
6. What do you think about the rise of India's software industry?

QUESTIONS FOR ENTREPRENEURS IN BANGALORE

Entrepreneurs in Bangalore were asked the same questions regarding educational and professional background and family history. The questions on Silicon Valley were reversed to reflect the entrepreneurs' presence in Bangalore under entrepreneurial history. Under the category "Nature of Business," I asked about the companies' ties to the U.S. software industry. The Open-Ended Questions did not include Question 4.

In addition, if the entrepreneurs were return migrants, the questionnaire included the immigrant history questions as well as the following:

1. When did you return to India?
2. Why did you decide to return to India?
3. Did you gain entrepreneurial experience in the United States?
4. How did your time overseas assist in your current entrepreneurial activities?

Endnote

1. The two women entrepreneurs were additionally asked about their experiences of being female entrepreneurs and what role gender played in their business practices.

References

AACI (Asian Americans for Community Involvement). (1993). *Qualified but...A report on the glass ceiling issues facing Asian Americans in Silicon Valley*. San Jose, CA: Asian Americans for Community Involvement.

Abbott, C. (1993). *The metropolitan frontier: Cities in the modern American West*. Tucson, AZ: University of Arizona Press.

Abraham, M., & Manning, S. (2004, August 14–17). *Business process outsourcing: The U.S. and India in the new global market*. Paper presented at the 2004 Annual Meeting of the American Sociological Association, San Francisco.

Adler, P., & Adler, P. (1994). Observational techniques. In N. Denzin & Y. Lincoln (Eds.), *Handbook of qualitative research* (pp. 372–392). Thousand Oaks, CA: Sage.

Aggarwal, A. (2003, February 10). Indo-U.S. economic ties: Growing stronger. Evalueserve White Paper. Retrieved April 2, 2008, from http://www.evalueserve.com/Media-And-Reports/WhitePapers.aspx#

Aiyar, S. (2004, November 21). IT can reduce regional disparities. *Times of India*. Retrieved April 2, 2008, from http://timesofindia.indiatimes.com/articleshow/929441.cms

Alba, R., & Nee, V. (1997). Rethinking assimilation theory for a new era of immigration. *International Journal of Migration Review, 31*, 826–874.

Albers, B. (2003). *On being a gun owner: Firearms and the cultivation of identity*. Unpublished Dissertation, Duke University.

Ananth, M. S., Babu, K. G., & Natarajan, R. (1989). *Database for brain drain: Institution-based study*. Department of Science and Technology, Government of India.

Anderson, B. (1991). *Imagined communities: Reflections on the origin and spread of nationalism*. New York and London: Verso.

Appadurai, A. (1993). Number in the colonial imagination. In C. Breckenridge & Peter van der Veer (Eds.), *Orientalism and the postcolonial predicament: Perspectives on South Asia* (pp. 314–339). Philadelphia: University of Pennsylvania Press.

Arora, A., Arunachalam, V. S., Asundi, J., & Fernandes, R. (2001). The Indian software services industry. *Research Policy, 30*(8), 1267–1287.

Asiaweek Special Report. (2000). Asia's best universities. Overall ranking: Science and technology schools. Retrieved April 2, 2008, from http://www.cgi.cnn.com/ASIANOW/asiaweek/features/universities2000/scitech.sci.overall.html

Baber, Z. (1996). *The Science of Empire*. Albany, NY: State University of New York Press.

Baker, S., & Kriplani, M. (1997, August 4). The global search for brainpower. *Business Week*, 46–50.

Barrett, G., Jones, T., & McEvoy, D. (1996). Ethnic minority business: Theoretical discourse in Britain and North America. *Urban Studies, 33*(4–5), 783–809.

Barrett, G., Jones, T., & McEvoy, D. (2001). Socio-economic and policy dimensions of the mixed embeddedness of ethnic minority business in Britain. *Journal of Ethnic and Migration Studies, 27*(2), 241–258.

Benson-Rea, M., & Rawlinson, S. (2003). Highly skilled and business migrants: Information processes and settlement outcomes. *International Migration, 41*(2), 59–79.

Bhagwati, J., & Partington, M. (1976). *Taxing the brain drain.* Amsterdam: North Holland.

Bhaneja, B. (1992). *Science and government: Nehru era.* New Delhi: National Publishing House.

Blum, J. (2000). Degradation without deskilling: Twenty-five years in the San Francisco shipyards. In Burawoy et al. (Eds.), *Global ethnography: Forces, connections and imaginations in a postmodern world* (pp. 106–136). Berkeley, CA: University of California Press.

Bonacich, E., & Modell, J. (1980). *The economic basis of ethnic solidarity: Small business in the Japanese American community*. Berkeley, CA: University of California Press.

Brewer, J. (1994). The ethnographic critique of ethnography. *Sociology, 28*, 231–244.

Brewer, J. (1996). An ethnography of crime in Belfast. Reply by John Brewer. *Journal of the Statistical and Social Inquiry Society of Ireland, 26*(3), 1–32.

Burawoy, M., Blum, J., George, S., Gille, Z., Gowan, T., Haney, L., Klawiter, M., Lopez, S., Riain, S., & Thayer, M. (2000). *Global ethnography: Forces, connections and imaginations in a postmodern world.* Berkeley, CA: University of California Press.

Burgers, J. (2000). A world of difference: Between the global and local. In D. Kalb, M. van der Land, R. Staring, B. van Steenbergen, & N. Wilterdink (Eds.), *The ends of globalization: Bringing society back in* (pp. 239–252). New York: Rowman & Littlefield Publishers.

Bylinsky, G. (1974, June). California's great breeding ground for industry. *Fortune*, 128–135.

Castells, M. (1996). *Rise of the network society*. Cambridge, MA: Blackwell.

Castells, M., & Portes, A. (1989). World underneath: The origins, dynamics, and effects of the informal economy. In A. Portes, M. Castells, & L. Benton (Eds.), *The informal economy: Studies in advanced and less developed countries* (pp. 11–37). Baltimore: Johns Hopkins University Press.

Chakravorty, B. (2006, October 14). Women rising. *Dataquest, 1*. Retrieved on April 2, 2008, from http://www.dqindia.ciol.com/content/top_stories/2006/106101406.asp

Cheng, L., & Yang, P. (1998). Global interaction, global inequality, and migration of the highly trained to the United States. *International Migration Review, 32*, 627–653.

Cho, W., & Lad, S. (2004). Subcontinental divide: Asian Indians and Asian American politics. *American Politics Research, 32*(3), 239–263.

Chopra, G., Manasseh, K., & Neal, C. (2004, June). *Sustaining India's services revolution: Access to foreign markets, domestic reform and international negotiations*. World Bank Report.

Cohn, B. (1987). The census, social structure and objectification in South Asia. In B. Cohn (Ed.), *An anthropologist among the historian and other essays* (pp. 224–254). Delhi and London: Oxford University Press.

Cornelius, W., Martin, P., Hollifield, J., & Tsuda, T. (2003). *Controlling immigration: A global perspective*. Stanford, CA: Stanford University Press.

D'Costa, A. (2003). Uneven and combined development: Understanding India's software exports. *World Development, 31*(1), 211–226.

Deb, S. (2004). *The IITians: The story of a remarkable Indian institution and how its alumni are reshaping the world.* New Delhi: Penguin Books.

Dedrick, J., & Kraemer, K. (1993). Information technology in India: Quest for self-reliance. *Asian Survey, 33*(5), 463–492.

Dicken, P. (2003). *Global shift: Transforming the world economy.* London: Sage Publications.

Dickson, B. (1996). Discussion of ethnography of crime in Belfast. *Journal of the Statistical and Social Inquiry Society of Ireland, 26*(3), 1–32.

Dillon, S. (2004, November 10). Foreign enrolments decline at US universities, surveys say. *New York Times.* Retrieved April 2, 2008, from http://www.nytimes.com/2004/12/21/national/21global.html?_r=&oref=slogin

Dirks, N. (1987). *The hollow crown: Ethnohistory of an Indian kingdom.* Cambridge, U.K.: Cambridge University Press.

Dossani, R. (2002). *Chinese and Indian engineers and their networks in Silicon Valley.* Working Paper, Asia-Pacific Research Center, Stanford University. Retrieved February 20, 2008, from http://cisac.stanford.edu/publications/12082/

Drezner, D. (2004). The outsourcing bogeyman. *Foreign Affairs, 83*(3), 1–6.

Dudley, K. (1994). The end of the line: Lost jobs, new lives in postindustrial America. Chicago: University of Chicago Press.

Dumont, L. (1980). *Homo hierarchicus: The caste system and its implications.* Chicago: University of Chicago Press.

Emerson, R., Fretz, R., & Shaw, L. (1995). *Writing ethnographic field notes*. Chicago: University of Chicago Press.

Eriksen, T. (2001). Between universalism and relativism: A critique of the Unesco concept of culture. In J. Cowan, M.-B. Dembour, & R. Wilson (Eds.), *Culture and rights: Anthropological perspectives* (pp. 127–148). Cambridge, U.K.: Cambridge University Press.

Evans, P. (2000). Fighting marginalization with transnational networks: Counter-hegemonic globalization? *Contemporary Sociology, 29*, 230–241.

FAIR (Federation for American Immigration Reform). (2003). Extended immigration data for North Carolina. Retrieved February 20, 2008, from http://www.fairus.org/site/PageServer?pagename=research_research8875

Fernandez Kelly, M. (1983). *For we are sold, I and my people: Women and industry in Mexico's frontier*. Albany, NY: State University of New York Press.

Findlay, A. (1995). Skilled transients: The invisible phenomenon? In R. Cohen (Ed.), *The Cambridge survey of world migration* (pp. 515–522). Cambridge, U.K.: Cambridge University Press.

Findlay, J. (1992). *Magic lands: Western cityscapes and American culture after 1940*. Berkeley, CA: University of California Press.

Foner, N. (1997). *What's new about transnationalism? New York immigrants today and at the turn of the century*. Paper for the conference on "Transnational Communities and the Political Economy of New York in the 1990s," New School for Social Research, New York.

Geertz, C. (1973). *Thick description: Toward an interpretive theory of cultures*. New York: Basic Books.

Gellner, E. (1964). *Thought and change*. London: Weidenfeld and Nicholson.

Gereffi, G. (1994). The international economy. In N. Smelser & R. Swedberg (Eds.), *The handbook of economic sociology* (pp. 206–233). Princeton, NJ: Princeton University Press/ Russell Sage Foundation.

Gibbs, W. (1994). Software's chronic crisis. *Scientific American, 271*(3), 86–95.

Giddens, A. (1971). *Capitalism and modern social theory: Analysis of the writngs of Marx, Durkheim and Max Weber*. Cambridge, U.K.: Cambridge University Press.

Gille, Z. (2001). Studying globalization critically, towards a new concept of site. *Cultural Studies-Critical Methodologies, 1*(3), 319–334.

Gille, Z., & Riain, S. (2002). Global ethnography. *Annual Review of Sociology, 28*, 271–95.

Glick Schiller, N., Basch, L., & Szanton-Blanc, C. (1992).Transnationalism: A new analytic framework for understanding migration. In N. Glick Schiller, L. Basch, & C. Szanton-Blanc (Eds.), *Toward a transnational perspective on migration* (pp. 1–24). New York: New York Academy of Sciences.

Goldman, M. (2001). The birth of a discipline: Producing authoritative green knowledge, World-Bank style. *Ethnography, 2*(2), 191–217.

Goldring, L. (1996). Gendered memory: Constructions of rurality among Mexican transnational migrants. In E. Melanie DuPuis & P. Vandergeest (Eds.), *Creating the countryside: The politics of rural and environmental discourse* (pp. 303–329). Philadelphia: Temple University Press.

Gowan, T. (2000). Excavating "globalization" from street level: Homeless men recycle their pasts. In Burawoy et al. (Eds.), *Global ethnography: Forces, connections and imaginations in a postmodern world* (pp. 74–105). Berkeley, CA: University of California Press.

Grasmuck, S., & Pessar, P. (1991). *Between two islands: Dominican international migration.* Berkeley, CA: University of California Press.

Gray, M., Golob, E., Markusen, A., & Ock Park, S. (1999). The four faces of Silicon Valley. In A. Markusen, Y-S. Lee, & S. DiGiovanna (Eds.), *Second tier cities* (pp. 291–310). Minneapolis, MN: University of Minnesota Press.

Griffin, P., & Chatham, R. (1958). Urban impact on agriculture in Santa Clara County, California. *Annals of the Association of American Geographers, 48*(September 1958), 195–208.

Guarnizo, L., Portes, A., & Haller, W. (2002). Transnational entrepreneurs: The emergence and determinants of an alternative form of immigrant adaptation. *American Sociological Review, 67*, 278–298.

Guarnizo, L., Portes, A., & Haller, W. (2003). Assimilation and transnationalism: Determinants of transnational political action among contemporary immigrants. *American Journal of Sociology, 108*(6), 1211–1248.

Guha, R. (1983). The prose of counter-insurgency. In R. Guha (Ed.), *Subaltern studies: Writings on South Asian history and society* (pp. 45–88). New Delhi and London: Oxford University Press.

Hall, S. (1993). Culture, community, nation. *Cultural Studies, 7*(3), 349–363.

Hammersley, M. (1989). *The dilemma of qualitative method.* London: Routledge.

Hammersley, M. (1990). *Reading ethnographic research.* London: Routledge.

Hammersley, M. (1992). *What's wrong with ethnography?* London: Routledge.

Hammersley, M., & Atkinson, P. (1983). *Ethnography: Principles in practice.* New York: Tavistock Publications.

Hanson, D. (1983). *The new alchemists: Silicon Valley and the microelectronics revolution.* Boston: Little, Brown.

Heeks, R. (1996). *India's software industry: State policy, liberalisation and industrial development.* New Delhi: Sage Publications.

Henry, N., McEwan, C., & Pollard, J. (2002). Globalization from below: Birmingham—postcolonial workshop of the world? *Area, 34*(2), 118–127.

Hoefler, D. (1971, January 11, 18, 25). Silicon Valley—USA. *Electronic News.*

Howard, P. (2002). Network ethnography and the hypermedia organization: New media, new organizations, new methods. *New Media and Society, 4*(4), 550–574.

Hum, T. (2001). The promises and dilemmas of immigrant ethnic economies. In M. Lopez-Garcia & D. Diaz (Eds.), *Asian and Latino immigrants in a restructuring economy: The metamorphosis of Los Angeles* (pp. 77–101). Stanford, CA: Stanford University Press.

Huntington, S. (1993). *The clash of civilizations and the remaking of the world order.* New York: Touchstone.

Immigration & Nationality Acts Amendments. (1965). Immigration and nationality act amendments of October 3, 1965 (79 Statutes-at-Large 911).

Immigration and Naturalization Service. (2001). *Statistical Yearbook*. U.S. Department of Justice.

Indiresan, P. (2003, February 7). IITs: Invaluable institutions. *The Hindu Business Line*. Retrieved April 2, 2008, from http://www.thehindubusinessline.com/2003/02/07/stories/2003020700070800.htm

Indiresan, P., & Nigam, N. (1993). The Indian Institutes of Technology: Excellence in peril. In P. Altbach & S. Chitnis (Eds.), *Higher education reform in India: Experience and perspectives*, 334–364. New Delhi and Newbury Park, CA: Sage Publications.

Iredale, R. (1999). The need to import skilled personnel. *International Migration Special Issue: Migration and Development, 37*(1), 89–123.

Iredale, R. (2001). The migration of professions: Theories and typologies. *International Migration, 39*(5), 7–26.

Itzigsohn, J., Dore, C., Hernandez, E., & Vazquez, O. (1999). Mapping Dominican transnationalism. *Ethnic and Racial Studies, 22*, 422–446.

Jensen, J. (1988). *Passage from India: Asian Indian immigrants in North America*. New Haven, CT: Yale University Press.

Joint India-U.S. Statement. (2000). India-U.S. relations: A vision for the 21st century. Embassy of India. Statement issued March 21 at New Delhi.

Kabra, K. (1977). Brain drain—the missing perspective. *Journal of Higher Education, 3*(2), 348–364.

Kalleberg, A. (2000). Nonstandard employment relations: Part-time, temporary and contract work. *American Sociological Review, 26*, 341–365.

Kandel, W., & Parrado, E. (2004). Hispanics in the American South and the transformation of the poultry industry. In D. Arreola (Ed.), *Hispanic spaces, Latino places: Community and cultural diversity in contemporary America* (pp. 255–276). Austin, TX: University of Texas Press.

Kanjanapan, W. (1995). The immigration of Asian professionals to the United States: 1988–1990. *International Migration Review, 29*(1), 7–32.

Kaplan, D. (1999). *The silicon boys and their valley of dreams.* New York: William Morrow and Company Inc.

Kenney, M. (2000). *Understanding Silicon Valley: The anatomy of an entrepreneurial region.* Stanford, CA: Stanford University Press.

Khadria, B. (1999). *The migration of knowledge workers: Second generation effects of India's brain drain.* New Delhi: Sage Publications.

Khadria, B. (2001). Shifting paradigms of globalization: The twenty-first century transition towards generics in skilled migration from India. *International Migration, 39*(5), 45–71.

Khadria, B. (2004a). Human resources in science and technology in India and the international mobility of highly skilled Indians. *OECD STI (Science, Technology and Industry) Working Paper Series,* (7).

Khadria, B. (2004b). Migration of highly skilled Indians: Case studies of IT and health professionals. *OECD STI (Science, Technology and Industry) Working Paper Series,* (6).

Khadria, B. (2007). Tracing the genesis of brain drain in India through its state policy and civil society. In N. Green & F. Weil

(Eds.), *Citizenship and those who leave: The politics of emigration and expatriation* (pp. 265–282). Champaigne, IL: University of Illinois Press.

Kloosterman, R., van der Leun, J., & Rath, J. (1999). Mixed embeddedness: (In)formal activities and immigrant businesses in the Netherlands. *International Journal of Urban and Regional Research, 23*(2), 252–266.

Kobayashi-Hilary, M. (2004). *Outsourcing to India: The offshore advantage*. New York: Springer-Verlag.

Kogut, B. (1985). Designing global strategies: Comparative and competitive value-added strategies. *Sloan Management Review, 26*(4), 15–28.

Konrad, R. (2003, August 15). Training Your Own Replacement. *CBS News*. Retrieved February 20, 2008, from http://www.cbsnews.com/stories/2003/08/14/tech/main568324.shtml

Koser, K., & Salt, J. (1997). The geography of highly skilled international migration. *International Journal of Population Geography, 3*, 285–303.

Kumar, D. (1995). *Science and the Raj: 1857–1905*. New Delhi: Oxford University Press.

Kuper, A. (1999). *Culture: The anthropologists' account*. Cambridge, MA: Harvard University Press.

Kyle, D. (1999). The Otavalo trade diaspora: Social capital and transnational entrepreneurship. *Ethnic and Racial Studies, 22*(2), 422–446.

Lakha, S. (1994). The new international division of labour and the Indian computer software industry. *Modern Asian Studies, 28*(2), 381–408.

Landolt, P., Autler, L., & Baires, S. (1999). From "Hermano Lejano" to "Hermano Major": The dialectics of Salvadoran transnationalism. *Ethnic and Racial Studies, 2*, 290–215.

Lateef, A. (1997). *Linking up with the global economy: A case study of the Bangalore software industry*. Discussion paper for the International Institute for Labour Studies, Geneva, Switzerland.

Lawson, V. (1992). Industrial subcontracting and employment forms in Latin America. *Progress in Human Geography, 16*, 1–23.

Lenn, M., & Campos, L. (eds.). (1996). *Globalisation of the professions and the quality imperative: Professional accreditation, certification, and licensure*. Madison, WI: Magna Publications.

Leslie, S. (2000). The biggest "angel" of them all: The military and the making of Silicon Valley. In M. Kenney (Ed.), *Understanding Silicon Valley: The anatomy of an entrepreneurial region* (pp. 48–67). Stanford, CA: Stanford University Press.

Leung, M. (2001). Get IT going: New ethnic Chinese business. The case of Taiwanese-owned computer firms in Hamburg. *Journal of Ethnic and Migration Studies, 27*(2), 277–294.

Lewis, M. (2000). *The new new thing: A Silicon Valley story*. New York: W. W. Norton.

Li, F., Findlay, A., Jowett, A., & Skeldon, R. (1996). Migrating to learn and learning to migrate: A study of the experiences and intentions of international student migrants. *International Journal of Population Geography, 2*, 51–67.

Light, I. (1984). Immigrant and ethnic enterprise in America. *Ethnic and Racial Studies, 7*, 195–216.

Light, I., & Bonacich, E. (1988). *Immigrant entrepreneurs: Koreans in Los Angeles, 1965–1982*. Berkeley, CA: University of California Press.

Light, I., & Gold, S. (2000). *Ethnic economies*. San Diego, CA: Academic Press.

Light, I., Kim, R., & Hum, C. (2000). Globalization, vacancy chains, or migration networks? Immigrant employment and income in greater Los Angeles. In D. Kalb, M. van der Land, R. Staring, B. van Steenbergen, & N. Wilterdink (Eds.), *The ends of globalization: Bringing society back in* (pp. 217–239). New York: Rowman & Littlefield Publishers.

Lim, L. (1980). Women workers in multinational corporations: The case of the electronics industry in Malaysia and Singapore. In K. Kumar (Ed.), *Transnational enterprises: Their impact on Third World societies and cultures* (pp. 109–136). Boulder, CO: Westview Press.

Logan, J., Alba, R., & McNulty, T. (1994). Ethnic economies in metropolitan regions: Miami and beyond. *Social Forces, 72*, 691–724.

Lowen, R. (1997). *Creating the Cold War university: The transformation of Stanford*. Berkeley, CA: University of California Press.

Ludden, D. (1993). Orientalist empiricism: Transformation of colonial knowledge. In C. Breckenridge & P. van der Veer (Eds.), *Orientalism and the postcolonial predicament: Perspectives on South Asia* (pp. 250–278). Philadelphia: University of Pennsylvania Press.

Luger, M., & Goldstein, H. (1991). *Technology in the garden: Research parks and regional economic development*. Chapel Hill, NC: University of North Carolina Press.

Macaulay, T. (1835). Minute of 2 February 1835 on Indian education. Retrieved February 20, 2008, from http://shelterbelt.com/ARCHIVE/macaulayminutes.html

Mahler, S. (2002). Theoretical and empirical contributions toward a research agenda for transnationalism. In M. Smith & L. Guarnizo (Eds.), *Transnationalism from below* (pp. 64–100). New Brunswick, NJ: Transaction Publishers.

Mahon, T. (1985). *Charged bodies: People, power and paradox in Silicon Valley*. New York; Scarborough, ON: New American Library.

Malecki, E. (1997). *Technology and economic development: The dynamics of local, regional and national competitiveness*. London and Boston: Addison Wesley Longman.

Malone, M. (2002). *The Valley of Heart's Delight: A Silicon Valley notebook*, 1963–2001. New York: John Wiley and Sons.

Markusen, A., Hall, P., Campbell, S., & Deitrick, S. (1991). *The rise of the gunbelt: The military remapping of industrial America*. New York: Oxford University Press.

Mato, D. (1997). On global and local agents and the social making of transnational identities and related agendas in "Latin" America. *Identities, 4*(2),167–212.

Matthews, G. (1999). "The Los Angeles of the North" San Jose's transition from fruit capital to high-tech metropolis. *Journal of Urban History, 25*(4), 459–496.

Meijering, L., & van Hoven, B. (2003). Imagining difference: The experience of "transnational" Indian IT professionals in Germany. *Area, 35*(2), 174–182.

Merchant, K. (2003, July 2). Services waves ripples outwards. *Financial Times*. Retrieved on April 7, 2008, from http://

search.ft.com/ftArticle?queryText=khozem+merchant+2003&page=1&y=0&drillDown=%2Bgacompanyname%3A%5E%22Tata+Consultancy+Services+Ltd%22%24&aje=false&x=0&id=030702001539&ct=0

Merton, R. (1987). Three fragments from a sociologist's notebook: Establishing the phenomena, specified ignorance, and strategic research materials. *Annual Review of Sociology, 13*, 1–28.

Mills, C. (1970) [1959]. *The sociological imagination.* Hardmondsworth, U.K.: Penguin.

Min, P., & Bozorgmehr, M. (2000). Immigrant entrepreneurship and business patterns: A comparison of Koreans and Iranians in Los Angeles. *International Migration Review, 34*(3), 707–738.

Ministry of Education and Culture, Government of India. (January 1980). Report of the review committee on foreign technical assistance received by the Indian Institutes of Technology and other academic institutions.

Mitra, P., & Schaffer, T. (2004, June 3). India: Expect turbulence. *South Asia Monitor*, 71. Center for Strategic & International Studies Newsletter. Retrieved April 7, 2008, from http://www.csis.org/media/csis/pubs/sam71.pdf

Mooney, P., & Neelakantan, S. (2004, October 8). No longer dreaming of America. *The Chronicle of Higher Education, 51*(7), 41.

Moore, B. (1966). *Social origins of dictatorship and democracy: Lord and peasant in the making of the modern world.* Boston: Beacon Press.

Morgan, G. (2001). Transnational communities and business systems. *Global Networks, 1*(2), 113–130.

Murali, K. (2003, February 1–14). The IIT story: Issues and concerns. *Frontline, 20*(3). Retrieved on April 7, 2008, from http://www.flonnet.com/fl2003/stories/20030214007506500.htm

Naroola, G. (2001). *The entrepreneurial connection: East meets West in Silicon Valley*. New Delhi: Tata McGraw Hill.

Nash, J., & Fernandez Kelly, M. (1983). *Women, men and the new international division of labor*. Albany, NY: State University of New York Press.

NASSCOM (National Association of Software and Services Companies). (2003). *IT industry fact sheet. Indian software and services industry: NASSCOM analysis.*

NASSCOM. (2004a). *The Indian ITES-BPO industry overview.*

NASSCOM. (2004b). *IT software and services market.*

Nolan, C. (2001, March 8). Clinton's India ties. *Wired.*

National Science Board. (2000) *Science and engineering indicators, Volumes 1 and 2.* Arlington, VA: National Science Foundation. Retrieved on April 7, 2008, from http://www.nsf.gov/statistics/seind00/pdfstart.htm

O'Mara, M. (2004). *Cities of knowledge: Cold War politics and the suburbanization of science*. Princeton, NJ: Princeton University Press.

Omvedt, G. (1993). *Reinventing revolution: New social movements and the socialist tradition in India.* Armonk, NY: ME Sharpe.

Ong, P., Bonacich, E., & Cheng, L. (1994). The political economy of capitalist restructuring and the new Asian immigration. In P. Ong, E. Bonacich, & L. Cheng (Eds.), *The new*

Asian immigration in Los Angeles and global restructuring (pp. 3–35). Philadelphia: Temple University Press.

Ong, P., Cheng, L., & Evans, L. (1992). Migration of highly educated Asians and global dynamics. *Asian and Pacific Migration Journal*, 1(3–4), 543–567.

Open Doors. (2004). *International students in the U.S.* Annual Report of the Institute of International Education. Retrieved on April 7, 2008, from http://opendoors.iienetwork.org/?p=50137

Open Doors. (2007). *International students in the U.S. rebounds*. Annual Report of the Institute of International Education. Retrieved on April 7, 2008, from http://opendoors.iienetwork.org/?p=113743

O'Riain, S. (2004). The politics of mobility in technology-driven commodity chains: Development coalitions in the Irish software industry. *International Journal of Urban and Regional Research, 28*(3), 642–663.

Parthasarathy, B. (2004). India's Silicon Valley or Silicon Valley's India? Socially embedding the computer software industry in Bangalore. *International Journal of Urban and Regional Research, 28*(3), 664–685.

Patibandla, M., & Petersen, B. (2002). The role of transnational corporations in the evolution of a high-tech industry: The case of India's software industry. *World Development, 30*(9), 1561–1577.

Phillips, M., & Stahl, C. (2001). International trade in higher education services in the Asia Pacific region: Trends and issues. *Asian and Pacific Migration Journal, 10*(2), 273–301.

PISA (Program for International Student Assessment). (2003). *OECD (Organization for Economic Cooperation and Development) Report*. Retrieved February 20, 2008, from http://www.pisa.oecd.org/dataoecd/1/60/34002216.pdf

Planning Commission, Government of India. (1956). *Second five year plan*. New Delhi, India. Retrieved February 20, 2008, from http://planningcommission.nic.in/plans/planrel/fiveyr/welcome.html

Portes, A. (2000). Globalization from below: The rise of transnational communities. In D. Kalb, M. van der Land, R. Staring, B. van Steenbergen, & N. Wilterdink (Eds.), *The ends of globalization: Bringing society back in* (pp. 253–270). New York: Rowman & Littlefield Publishers.

Portes, A., (Ed.). (1995). *The economic sociology of immigration: Essays on network, ethnicity and entrepreneurship*. New York: Russell Sage Foundation.

Portes, A., & Bach, R. (1985). *Latin journey: Cuban and Mexican immigrants in the United States*. Berkeley and Los Angeles, CA: University of California Press.

Portes, A., Guarnizo, L., & Landolt, P. (1999). The study of transnationalism: Pitfalls and promise of an emergent field. *Ethnic and Racial Studies, 22*(2), 217–237.

Portes, A., & Rumbaut, R. (1996). *Immigrant America*. Berkeley, CA: University of California Press.

Portes, A., & Zhou, M. (1992). Gaining the upper hand: Economic mobility among immigrant and domestic minorities. *Ethnic and Racial Studies, 15*(4), 491–522.

Prieto, Y. (1992). Cuban women in New Jersey: Gender relations and change. In D. Gabaccia (Ed.), *Seeking common ground: Multidisciplinary studies of immigrant women in the United States* (pp. 185–202). Westford, CT: Westwood Press.

Prus, R. (1996). *Symbolic interaction and ethnographic research: Intersubjectivity and the study of human lived experience*. Albany, NY: State University of New York.

Rahman, A., Bhargava, R., Qureshi, M., & Pruthi, S. (1973). *Science and technology in India.* New Delhi: Indian Council for Cultural Relations.

Rai, S. (2004, November 4). An industry in India cheers Bush's victory. *New York Times*. Retrieved April 7, 2008, from http://www.nytimes.com/2004/11/04/business/worldbusiness/04outsource.html?_r=1&oref=slogin

Raj, K. (2000). Colonial encounters and the forging of new knowledge and national identities: Great Britain and India, 1760–1850. In R. McLeod (Ed.), *Nature and empire: Science and the colonial enterprise* (pp. 119–134). Ithaca, NY: Osiris.

Rajghatta, C. (2003, January 18). IITs better than U.S. institutes, says CBS. *Times of India*. Retrieved February 20, 2008, from http://timesofindia.indiatimes.com/articleshow/34701392.cms

Ramamurti, R. (2004). Developing countries and MNEs: Extending and ecnriching the research agenda. *Journal of International Business Studies, 35*(4), 277–283.

Reinharz, S. (1992). *Feminist methods in social research.* New York: Oxford University Press.

Rogers, E., & Larsen, J. (1984). *Silicon Valley fever: Growth of high-technology culture*. New York: Basic Books.

Safa, H. (1995). *The myth of the male breadwinner: Women and industrialization in the Caribbean*. Boulder, CO: Westview Press.

Salkever, A. (1999, December 6). Technical Sutra. *Salon*. Retrieved February 20, 2008, from http://www.salon.com/books/it/1999/12/06/indian/

Sassen, S. (1998). *Globalization and its discontents*. New York: The New Press.

Savage, M., Warde, A., & Ward, K. (2003). *Urban sociology, capitalism and modernity*. New York: Palgrave MacMillan.

Saxenian, A. (1994). *Regional advantage: Culture and competition in Silicon Valley and Route 128*. Cambridge, MA: Harvard University Press.

Saxenian, A. (1999). *Silicon Valley's new immigrant entrepreneurs*. San Francisco: Public Policy Institute of California.

Saxenian, A. (2002). *Local and global networks of immigrant professionals in Silicon Valley*. San Francisco: Public Policy Institute of California.

Scheper-Hughes, N. (1992). *Death without weeping: The violence of everyday life in Brazil*. Berkeley, CA: University of California Press.

Schware, R. (1992). Software industry entry strategies for developing countries: A "walking on two legs" proposition. *World Development, 20*(2), 143–164.

Sheth, D., & Mahajan, G. (1999). *Minority identities and the nation-state*. Delhi: Oxford University Press.

Shore, C. (2000). *Building Europe: Cultural politics of European integration*. London and New York: Routledge.

Singh, B. (1988). *Jawaharlal Nehru on science and society: A collection of his writings and speeches*. New Delhi: Nehru Memorial Museum and Library.

Sipahimani, V. (2000). Financing of elementary education in India in the 1990s. *World Bank Report*. South Asia Education Sector, Technical Working Paper No. 2. New Delhi, India.

Sklair, L. (1995). *Sociology of the global system*. Baltimore: Johns Hopkins University Press.

Sklair, L. (2001). *The transnational capitalist class*. Oxford: Blackwell.

Slaughter, M. (2004a, March 26). Globalization and employment by U.S. multinationals: A framework and facts. *Daily Tax Report, Bureau of National Affairs*.

Slaughter, M. (2004b). Insourcing jobs: Making the global economy work for America. *Organization for International Investment Working Paper Series*. Retrieved February 20, 2008, from http://www.ofii.org/insourcing/insourcing_study.pdf

Smith, M. (2001). *Transnational urbanism: Locating globalization*. Oxford: Blackwell.

Smith, M., & Guarnizo, L. (Eds.). (2002). *Transnationalism from below*. New Brunswick: Transaction Publishers.

Spradley, J. (1980). *Participant observation*. New York: Holt, Rinehart, and Winston.

Stanley, L. (1990). Doing ethnography, writing ethnography. *Sociology, 24*, 617–628.

Staring, R. (2000). Flows of people: Globalization, migration and transnational communities. In D. Kalb, M. van der Land, R. Staring, B. van Steenbgergen, & N. Wilterdink (Eds.), *The ends of globalization: Bringing society back in* (pp. 203–216). New York: Rowman & Littlefield Publishers.

Stiglitz, J. (2003). *Globalization and its discontents*. New York: W. W. Norton & Company.

Sturgeon, T. (2000). How Silicon Valley came to be. In M. Kenney (Ed.), *Understanding Silicon Valley: The anatomy of an entrepreneurial region* (pp. 15–47). Stanford, CA: Stanford University Press.

Suchman, M. (2000). Dealmakers and counselors: Law firms as intermediaries in the development of Silicon Valley. In M. Kenney (Ed.), *Understanding Silicon Valley: The anatomy of an entrepreneurial region* (pp. 71–97). Stanford, CA: Stanford University Press.

Sukhatme, S., & Mahadevan, I. (1987). *Pilot study on magnitude and nature of the brain drain of graduates of the Indian Institute of Technology, Bombay*. Department of Science and Technology, Government of India.

Takaki, R. (1989). *Strangers from a different shore: A history of Asian Americans*. New York: Penguin Books.

Teixeira, C. (2001). Community resources and opportunities in ethnic economies: A case study of Portuguese and black entrepreneurs in Toronto. *Urban Studies, 38*(11), 2055–2078.

Terman, F. (1947). *Dean's report, School of Engineering, 1946–47*. S. E. Wallace-Sterling Papers. Stanford University Special Collections 39/School of Engineering.

Terman, F. (1978). Interviews conducted by A. L. Norberg, Charles Suskind and Rodger Hahn, History of Science and Technology Project, Bancroft Library, University of California Library.

Touraine, A. (1988). *Return of the actor: Social theory in postindustrial society*. Minneapolis, MN: University of Minnesota Press.

UNDP (United Nations Development Programme). (2001). Making new technologies work for human development. Retrieved on April 7, 2008, from http://hdr.undp.org/en/media/completenew1.pdf

U.S. Census Data. (Various years).

U.S. Citizenship and Immigration Services. (2003). H-1B frequently asked questions. (last modified 10/30/2003).

Valenzuela, A. (2003). Day-labor work. *Annual Review of Sociology, 29*(1), 661–692.

Varma, R., & Rogers, E. (2004, December). Indian cyber workers in the U.S. *Economic and Political Weekly, 39*(52), 5645–5652.

Vashistha, A., & Vashistha, A. (2004). *The offshore nation: Rise of services globalization*. San Ramon, CA: neoIT.com. Retrieved February 20, 2008, from http://www.neoit.com/gen/knowledgecenter/articles/OIv2i8_June_04_Excerpt_from_The_OffshoreNation.pdf

Vertovec, S. (1996). Berlin multikulti: Germany, "foreigners" and "world-openness." *New Community, 22*, 381–399.

Vertovec, S. (1999). Conceiving and researching transnationalism. *Ethnic and Racial Studies, 22*(2), 447–462.

Viswanathan, S. (1985). *Organizing for science: The making of an industrial laboratory*. Delhi: Oxford University Press.

Wagner, R. (1981). *The invention of culture*. Chicago: University of Chicago Press.

Waldinger, R. (2001). *Strangers at the gates: New immigrants in urban America*. Berkeley, CA: University of California Press.

Waldinger, R., Aldrich, H., & Ward, R. (1990). *Ethnic entrepreneurs: Immigrant business in industrial societies*. Newbury Park, CA: Sage Publications.

Waldman, A. (2004a, July 24). Indians go home but don't leave U.S. behind. *New York Times*. Retrieved on April 7, 2008, from http://query.nytimes.com/gst/fullpage.html?res=9405E3D8173DF937A15754C0A9629C8B63&st=cse&sq=amy+waldman&scp=18

Waldman, A. (2004b, July 17). A young American outsources himself to India. *New York Times*. Retrieved on April 7, 2008, from http://query.nytimes.com/gst/fullpage.html?res=9C07EED6153AF934A25754C0A9629C8B63&st=cse&sq=amy+waldman+young+american+outsources+himself&scp=1

Weber, M. (1946). India: The Brahman and the castes. In H. Gerth & C. Mills (Eds.), *From Max Weber: Essays in sociology* (pp. 396–415). New York: Oxford University Press.

Werbner, P. (1990). *The migration process: Capital, gifts and offerings among British Pakistanis*. New York: Berg.

Westwood, S., & Bhachu, P. (eds). (1988). *Enterprising women: Ethnicity, economy and gender relations*. London and New York: Routledge.

Wheeler, D. (2002, November 22). Boom times for border crossing. *The Chronicle of Higher Education, 49*(13), 62.

Wolf, D. (1992). *Factory daughters: Gender, household dynamics and rural industrialization in Java*. Berkeley, CA: University of California Press.

Wolfe, T. (1983, December). The tinkerings of Robert Noyce: How the sun rose on the Silicon Valley. *Esquire*, 346–374.

World Bank Country Report: India. (2003, August). Retrieved on April 7, 2008, from http://www.hrw.org/wr2k3/asia6.html

Xiang, B. (2001). Structuration of Indian information technology professionals' migration to Australia: An ethnographic study. *International Migration, 39*(5), 73–90.

Yoon, I-J. (1997). *On my own: Korean businesses and race relations in America*. Chicago: University of Chicago Press.

Young, G. (1957). *Macaulay, prose and poetry*. Cambridge, MA: Harvard University Press.

Index

www.ingramcontent.com/pod-product-compliance
Lightning Source LLC
LaVergne TN
LVHW101442200926
848483LV00001B/23
* 9 7 8 1 6 0 4 9 7 5 2 7 7 *